STUDENT SOLUTION MANUAL FOR MATHEMATICAL INTEREST THEORY

Leslie Jane Federer Vaaler

Department of Mathematics
The University of Texas at Austin

Upper Saddle River, New Jersey 07458

Library of Congress Cataloging-in-Publication Data on file.

Student Solution Manual for Mathematical Interest Theory
1st ed./Leslie Jane Federer Vaaler
p. cm.
Includes references and index.
ISBN 0-13-238922-3
1. Numerical Analysis

Vice President and Editorial Director, ECS: *Marcia J. Horton*
Senior Editor: *Holly Stark*
Editorial Assistant: *Jennifer Lonschein*
Managing Editor: *David A. George*
Production Editor: *Craig Little*
Director of Creative Services: *Paul Belfanti*
Art Director: *Jayne Conte*
Cover Designer: *Bruce Kenselaar*
Marketing Manager: *Alexis Heydt-Long*
Manufacturing Buyer: *Lisa McDowell*
Senior Marketing Manager: *Tim Galligan*

Pearson Prentice Hall
Pearson Education, Inc.
Upper Saddle River, New Jersey 07458

Printed in the United States of America
10 9 8 7 6 5 4 3 2 1

ISBN 0-13-238922-3

Pearson Education Ltd., *London*
Pearson Education Australia Pty. Ltd., *Sydney*
Pearson Education Singapore, Pte. Ltd.
Pearson Education North Asia Ltd., *Hong Kong*
Pearson Education Canada, Inc., *Toronto*
Pearson Educación de Mexico, S.A. de C.V.
Pearson Education – Japan, *Tokyo*
Pearson Education Malaysia, Pte. Ltd.
Pearson Education, Inc., *Upper Saddle River, New Jersey*

Contents

Preface

About this manual

This manual is written to accompany *Mathematical Interest Theory* by James W. Daniel and Leslie Jane Federer Vaaler. It includes detailed solutions to the odd-numbered problems. There are solutions to 239 problems, and sometimes more than one way to reach the answer is presented . In keeping with the presentation of the text, calculator discussion for the Texas Instruments BAII Plus or BAII Plus Professional calculators is typeset in a different font from the rest of the text (the sans serif font).

Acknowledgements

In the preface to the text, the authors thanked their student accuracy checkers Carl Gillette, Karen Kimberly, and Gagan Nanda. Their duties included sharing their solutions to the problems with the authors, so it is appropriate to thank them again here.

The author also wishes to express her appreciation to the graduate students who have served as teaching assistants for her interest theory courses: thanks to Darice Chang, Miriam Fisk, Emma Fong, Anne Miller, Ana Neira, Rebecca Armon, and Kristen Tanaka.

Thanks also to the many students who came to office hours. You asked me to explain how to solve many of these problems and made me aware of common pitfalls and areas of confusuion. This manual is surely a better book because of you!

Finally, to my family, I again offer my gratitude. You, along with my pastels, add color and fun.

Contacting the author

If you note errors of any sort, kindly send an e-mail message reporting them; the author's e-mail address is `lvaaler@math.utexas.edu`. The author would also appreciate receiving any other comments you wish to make.

Leslie Jane Federer Vaaler

C H A P T E R 0

An introduction to the Texas Instruments BA II Plus

(0.3) BA II Plus calculator Basics

(1) To set to 9–formatting, key [2ND] [FORMAT] [9] [ENTER] [2ND] [QUIT]. Compute $(3.264)(1.0825)+(4.67)(1.065)$ by pressing

[(] [3] [•] [2] [6] [4] [×] [1] [•] [0] [8] [2] [5] [)] [+] [(] [4] [•] [6] [7] [×] [1] [•] [0] [6] [5] [)] [=].

At this point, the calculator will display "8.50683". Rounding 8.50683 to the nearest hundredth, you obtain 8.51, since 8.50683 is a number between 8.505 and 8.515. If you have 2–formatting and press the above string of keystrokes to compute $(3.264)(1.0825)+(4.67)(1.065)$, the display will show "8.51". On the other hand, if you have 2-formatting and push

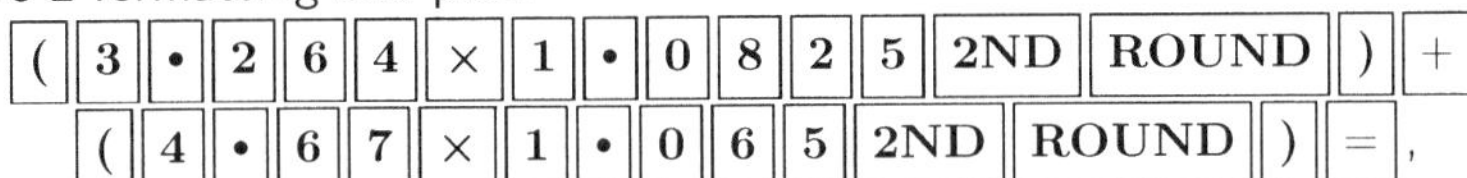

you will be adding the two rounded products, 3.53 plus 4.97; the result is 8.50.

(3) (a) You may compute $500(1.04)^5+800(1.03)^4$ by keying

[(] [5] [0] [0] [×] [(] [1] [•] [0] [4] [y^x] [5] [)] [=] [)] [+] [(] [8] [0] [0] [×] [(] [1] [•] [0] [3] [y^x] [4] [)] [)] [=];

this yields 1,508.733499. If you prefer not to use parentheses, you can also obtain this answer by first keying [1] [•] [0] [4] [y^x] [5] [×] [5] [0] [0] [=] to compute $500(1.04)^5$ and storing the resulting 608.3264512 in you favorite register m (where m an integer of your choosing between 0 and 9); to do this, key [STO] [m]. Next, calculate $800(1.03)^4$ by pressing [1] [•] [0] [3] [y^x] [4] [×] [8] [0] [0] [=] and finally key [+] [RCL] [m] [=] to obtain the desired sum. Again, you should obtain 1,508.733499 . To the nearest ten-thousandth, as specified in the problem, the answer is 1,508.7335.

(b) The ratio $\frac{\ln(32{,}546)}{\ln(1.05)}$ may be calculated by computing [3] [2] [5] [4] [6] [LN] [÷] [(] [1] [•] [0] [5] [LN] [)] [=]. The result is 212.9611558. To the nearest ten-thousandth, the answer is 212.9612.

As in part (a), you may store an intermediate result, which you recall later, if you prefer not to use parentheses.

(c) Keying [1] [•] [0] [5] [2ND] [e^x] [−] [1] [=] gives $e^{1.05}-1\approx 1.857651118$ so, to the nearest ten-thousandth, the answer is 1.8577.

CHAPTER 1

The growth of money

(1.3) Accumulation and amount functions

(1) In order to determine K, use the property $A_K(0) = K$. We have $A_K(0) = \frac{1{,}000}{100-0} = 10$, so $K = 10$. Therefore, $a(20) = \frac{A_K(20)}{K} = \frac{1{,}000/80}{10} = 1.25$.

(3) Firstly, observe that $a(0) = 0$ forces $\beta = 1$. Secondly, $.02 = i_1 = \frac{a(1)-a(0)}{a(0)} = a(1) - 1$ so $1.02 = \alpha(1^2) + .01(1) + 1 = \alpha + .01$, and $\alpha = .01$. Therefore, $a(t) = .01t^2 + .01t + 1$, and $i_4 = \frac{a(4)-a(3)}{a(3)} = \frac{1.2-1.12}{1.12} \approx .071428571 \approx 7.14285\%$.

(5) The amount of interest earned from time 0 to n is the sum of the amount of interest earned in the firtst n time periods, namely

$$1 + 2 + \cdots + n = \frac{1}{2}n(n+1);$$

this equality may be obtained by noting that

$$\begin{aligned}
1 + 2 + \cdots + n &= \frac{1}{2}[(1 + 2 + \cdots + n) + (1 + 2 + \cdots + (n-1) + n)] \\
&= \frac{1}{2}[(1 + 2 + \cdots + n) + (n + (n-1) + \cdots + 2 + 1)] \\
&= \frac{1}{2}[(1 + n) + (2 + (n-1)) + \cdots + ((n-1) + 2) + (n+1)] \\
&= \frac{1}{2}[(n+1) + (n+1) + \cdots + (n+1) + (n+1)] = \frac{1}{2}n(n+1).
\end{aligned}$$

(7) Note that $A_K(n-1) = 3(n-1)^2 + 2(n-1) + 800 = [3n^2 - 6n + 3] + [2n - 2] + 800 = 3n^2 + (-6+2)n + 800 = 3n^2 - 4n + 801$. Therefore,

$$\begin{aligned}
i_n &= \frac{A_K(n) - A_K(n-1)}{A_K(n-1)} = \frac{[3n^2 + 2n + 800] - [3n^2 - 4n + 801]}{3n^2 - 4n + 801} \\
&= \frac{6n - 1}{3n^2 - 4n + 801}.
\end{aligned}$$

If you know calculus, you may thus establish that $\{i_n\}$ is decreasing for $n \geq 17$ by showing that the real-values function $f(x) = \frac{6x-1}{3x^2-4x+801}$ has derivative $f'(x) < 0$ for $x \geq 17$. But,

$$\begin{aligned}
f'(x) &= \frac{6(3x^2 - 4x + 801) - (6x-1)(6x-4)}{(3x^2 - 4x + 801)^2} = \frac{(18x^2 - 24x + 4{,}806) - (36x^2 - 30x + 4)}{(3x^2 - 4x + 801)^2} \\
&= \frac{-18x^2 + 6x + 4{,}802}{(3x^2 - 4x + 801)^2}
\end{aligned}$$

Therefore, $f'(x)$ is negative whenever $-18x^2 + 6x + 4{,}802 < 0$. The quadratic formula may be used to find the roots of $-18x^2 + 6x + 4{,}802 = 0$. One root is negative and the other is $\frac{-6-\sqrt{6^2-4(-18)(4{,}802)}}{-2(18)} \approx 16.5$. For x larger than this positive root, so for $x \geq 17$, $f'(x)$ is negative.

If you do not know calculus, you may solve the problem by showing that for $n \geq 17$, the ratio

$$\frac{i_{n+1}}{i_n} = \frac{6(n+1)-1}{3(n+1)^2-4(n+1)+801}\bigg/\frac{6n-1}{3n^2-4n+801}$$

$$= \frac{(6n+5)(3n^2-4n+801)}{(6n-1)(3n^2+2n+800)}$$

$$= \frac{18n^3+9n^2+4{,}786n+4{,}005}{18n^3-9n^2+4{,}798n-800}$$

is less than 1. This is equivalent to establishing that for $n \geq 17$, the inequality

$$18n^3-9n^2+4{,}786n+4{,}005 < 18n^3+9n^2+4{,}798n-800$$

is true. But this inequality is equivalent to $18n^2-12n-4{,}805$ being positive. Since the largest root of the quadratic equation $18n^2-12n-4,805=0$ is $\frac{12+\sqrt{(-12)^2-4(18)(-4{,}805)}}{(2)(18)} \approx 16.67516788$, $18n^2-12n-4{,}805$ is indeed positive for $n \geq 17$. So, $\frac{i_{n+1}}{i_n} < 1$ for $n \geq 17$.

(1.4) Simple interest

(1) We have $A_{\$1{,}000}(t) = \$1{,}000(1+.05t)$, so $A_{\$1{,}000}(4) = \$1{,}200$ and $A_{\$1{,}000}(3) = \$1{,}150$. The amount of interest earned in the fourth year is $A_{\$1{,}000}(4) - A_{\$1{,}000}(3) = \$1{,}200 - \$1{,}150 = \$50$.

Alternatively, and more simply, with simple interest and a single investment of capital, the amount of interest is the same each year; it is the product of the amount invested and the annual rate of simple interest: In this case, the amount of interest earned each year is $\$1{,}000 \times .05 = \$50$. The balance at the end of the fourth year is $A_{\$1{,}000}(4) = \$1{,}000 + 4(\$50) = \$1{,}200$; the original \$1,000 had \$50 interest added for each of four years.

(3) The number of months elapsed is twelve times the number of years elapsed, since we view a month as $\frac{1}{12}$-th of a year. With simple interest , the amount of interest is given by the product Krt of the amount invested, the rate of simple interest and the time. Therefore, in order to to compensate for the time being multiplied by a factor of $\frac{1}{12}$, we must multiply the rate of simple interest by 12. Therefore, the annual rate of simple interest is $12 \times .5\% = 6\%$.

(5) We are given that $\$1{,}320 = 1{,}200(1+rT)$ where r is the rate of simple interest for this problem. So, $rT = \frac{\$1{,}320}{\$1{,}200} - 1 = .1$. We are asked to calculate $\$500[1+r(2T)]$. Since $r = .1$, it is equal to $\$500(1.2) = \$600$.

(7) Albert Einstein had many accomplishments, but his date of birth still had to be before his death. In fact, he was born on March 14, 1879, and he lived until April 18, 1955. We divide the interval between March 14, 1879 and April 18, 1955 into three subintervals, namely interval 1 from March 14, 1879 to December 31, 1949, interval 2 from December 31, 1949 to January 1, 1950, and interval 3 from January 1, 1950 to April 18, 1955. Note that the length of interval 2 was just one day.

Let interval 4 designate the interval from from March 14, 1979 to December 31, 2049; this interval is the interval precisely one hundred years after interval 1, and interval 1 has one fewer day than interval 4 since 1900 was *not* a leap year (since 1900 is divisible by 100 but not by 400) while 2000 was a leap year. We introduced interval 4 bacause its length may be calculated using the **Date worksheet**.

The number of days that Einstein lived was

$$\begin{aligned}&(\#\text{ days in interval 4 - 1}) + (\#\text{days in interval 2 }) + (\#\text{ days in interval 3})\\ &= (\#\text{ days in interval 3}) + (\#\text{ days in interval 4}),\end{aligned}$$

a sum of interval lengths that may each be calculated using the **Date worksheet**.

To calculate the number of days in interval 3, first key

2ND **DATE** **1** **•** **0** **1** **0** **0** **ENTER** **↓** **4** **•** **1** **8** **0** **5** **ENTER** ;

this will result in "DT1 = 1 – 01 – 2000" and "DT2 = 4 – 18 – 2005". Next key **↓** **↓**. If "ACT" is then displayed, find the number of days in interval 3 by keying **↑** **CPT**, while if the display shows "360", press **2ND** **SET** **↑** **CPT**. This should result in the display "DBD = 1,933", and there are 1,933 days in interval 3.

Similarly, using "DT1 = 3 – 14 – 79", "DT2 = 12 – 31 – 49", and "ACT", we find that interval 4 comprises 25,860 days. So, Einstein lived for 1,933 + 25,860 = 27,793 days.

(1.5) Compound interest

(1) We are given $A_K(t) = 2,200(1.04)^t$, and we wish to solve $A_K(T) = 8,000$. So, we need to find T such that $(1.04)^T = \frac{8,000}{2,200}$. To accomplish this, we take natural logarithms of each side of the equation, finding $\ln(1.04)^T = \ln\left(\frac{8,000}{2,200}\right)$. So, $T\ln(1.04) = \ln\left(\frac{8,000}{2,200}\right)$, and $T = \frac{\ln\left(\frac{8,000}{2,200}\right)}{\ln(1.04)} \approx 32.91587729 \approx 32.91588$.

(3) We are given that $(1+i)^9 = 2$. Therefore, $i = 2^{\frac{1}{9}} - 1 \approx .080059739 \approx 8.00597\%$.

(5) If we denote the annual interest rate by i, then we are given that $2 = (1+i)^\alpha$, $10 = (1+i)^\gamma$, and $5 = 12(1+i)^n$. It follows that $(1+i)^n = \frac{5}{12} = \frac{10}{24} = \frac{2^3 3}{10} = \frac{(1+i)^{3\alpha}(1+i)^\beta}{(1+i)^\gamma} = (1+i)^{3\alpha+\beta-\gamma}$. The function $f(x) = (1+i)^x$ is an increasing function, so it follows that $n = 3\alpha + \beta - \gamma$. We therefore have $n = a\alpha + b\beta + c\gamma$ with $a = 3$, $b = 1$, and $c = -1$.

(7) We need to find i such that $(1+i)^{14} = (1.05)^8(1.006)^{12\times 6}$; the reason we have the exponent 12×6 is that we were given a monthly interest rate during the last six years, and six years consists of 12×6 months. So, $i = [(1.05)^8(1.006)^{72}]^{\frac{1}{14}} - 1 \approx .060398768 \approx 6.03988\%$.

(9) The balance at the end of the first four years was $\$K[1 + (.025)4] = \$1.1K$, and if $t > 4$, the balance t years after the initial deposit of $K is $\$1.1K(1.05)^{t-4}$. We wish to find t so that this is equal to $\$3K$. Equivalently, we seek t with $1.1(1.05)^{t-4} = 3$. Therefore, $t - 4 = \frac{\ln(3/1.1)}{\ln(1.05)} \approx 20.56361412$ and $t \approx 24.56361412$. Starting in 1963, the balance triples in about 24.56361 years.

(11) (a) Applying the equation $a(s+t) = a(s)a(t)$ with $t = h$, we have $a(s+h) = a(s)a(h)$. Therefore,

$$\begin{aligned}a'(s) &= \lim_{h\to 0}\frac{a(s+h) - a(s)}{h} = \lim_{h\to 0}\frac{a(s)a(h) - a(s)}{h}\\ &= \lim_{h\to 0} a(s)\left[\frac{a(h)-1}{h}\right]\\ &= a(s)\lim_{h\to 0}\frac{a(h)-1}{h}.\end{aligned}$$

(b) Using the result from part (a) and the definition of the derivative $a'(0)$, we have

$$a'(s) = a(s)\left(\lim_{h\to 0}\frac{a(h)-1}{h}\right) = a(s)\left(\lim_{h\to 0}\frac{a(0+h)-1}{h}\right) = a(s)a'(0).$$

(c) Observe that $\frac{d}{ds}\ln a(s) = \frac{a'(s)}{a(s)}$ and, [from part (b)] $\frac{a'(s)}{a(s)} = a'(0)$. Therefore,

$$\int_0^t \frac{d}{ds}\ln a(s)ds = \int_0^t \frac{a'(s)}{a(s)}ds = \int_0^t a'(0)ds = a'(0)\int_0^t ds = a'(0)t.$$

(d) It follows from the Fundamental Theorem of Calculus and the result of part (c) that $\ln a(t) - \ln a(0) = a'(0)t$. But $\ln a(0) = \ln 1 = 0$, so $\ln a(t) = \ln a(t) - \ln a(0) = a'(0)t$.

(e) Recall that $a(1) = 1+i$. Therefore, applying the result of part (d) with $t = 1$, we have $\ln(1+i) = \ln a(1) = a'(0)\times 1 = a'(0)$.

(f) Combining the results of parts (d) and (e) and using an important property of logarithms , we find $\ln a(t) = a'(0)t = [\ln(1+i)]\,t = t\ln(1+i) = \ln(1+i)^t$.

(1.6) Effective discount rates/ Interest in advance

(1) Antonio gets the use of an extra $\$3,000 - .08(\$3,000) = \$3,000 - \$240 = \$2,760$.

(3) We are given that $\$1,320 = \$1,450 - \$1,450D = \$1,450(1-D)$. Therefore, $D = 1 - \frac{1,320}{1,450} \approx .089655172 \approx 8.96552\%$. Moreover, $I = (1-D)^{-1} - 1 = \frac{1,450}{1,320} \approx .098484848 \approx 9.84848\%$.

(5) Note that $1.2 = 1 + i_{[2,4.5]} = (1+i)^{4-2.5}$. So, $1 + i = (1.2)^{\frac{1}{2.5}} = (1.2)^{.4}$. Therefore, $d_{[1,3]} = 1 - (1 + i_{[1,3]})^{-1} = 1 - (1+i)^{-2} = 1 - (1.2)^{-.8} \approx .135718926 \approx 13.57189\%$.

(1.7) Discount functions/ The time value of money

(1) The money is invested at $t = 3$, but the accumulation function gives the growth of 1 invested at $t = 0$. Therefore, we first need to find the amount of money you would need to invest at $t = 0$ in order to have \$3,200 at $t = 8$; you would need $\frac{\$3,200}{a(8)} = \frac{\$3,200}{1.4}$. But $\frac{\$3,200}{1.4}$ deposited at $t = 0$ would grow to $\left(\frac{\$3,200}{1.4}\right)a(3) = \left(\frac{\$3,200}{1.4}\right)(1.15) \approx \$2,628.57$.

(3) Each year, the value of the home grows by a factor of 1.065. It cost \$156,000 on July 31, 2002, so its price P on July 31, 1998 satisfied $P(1.065)^4 = \$156,000$. Thus, $P = \$156,000(1.065)^{-4} \approx \$121,262.40$.

(5) We need to bring the \$5,000 back for ten years, using the appropriate discount factor for each year. Therefore, the present value is $\$5,000(1.04)^{-3}(1.05)^{-2}(1.055)^{-5} \approx \$3,084.814759 \approx \$3,084.81$.

(7) The present value of the first option is $6,000 + \$5,940(1+i)^{-1}$, and the second option has present value $\$12,000(1+i)^{-\frac{1}{2}}$. Therefore,

$$\$6,000 + \$5,940(1+i)^{-1} = \$12,000(1+i)^{-\frac{1}{2}}.$$

But this is equivalent to the equation

$$6,000(1+i) - \$12,000(1+i)^{\frac{1}{2}} + \$5,940 = 0.$$

Set $X = \$12{,}000(1+i)^{\frac{1}{2}}$. Then,

$$6{,}000X^2 - \$12{,}000X + \$5{,}940 = 0$$

and the quadratic formula gives

$$X = \frac{12{,}000 \pm \sqrt{(12{,}000)^2 - 4(5{,}940)(6{,}000)}}{2(6{,}000)} = \frac{12{,}000 \pm 1{,}200}{12{,}000} = 1 \pm .1.$$

So, $1 + i = X^2$ must either be equal to $1.1^2 = 1.21$ or to $.9.^2 = .81$ Assuming the loan is made at a positive rate of interest, it follows that $i = .21 = 21\%$.

(1.8) Simple discount

(1) Note that \$3,460 to be paid at $t = 9$ has a time $t = 0$ value of $\$3{,}460v(9) = \$3{,}460[1 - 9(.05)] = \$1{,}903$. Bringing this forward to time $t = 4$, we find a value of $\$1{,}903a(4) = \$1{,}903\frac{1}{1-4(.05)} = \$2{,}378.75$.

(3) The specified investment amount \$1,000 is not needed to do the problem. All we need to do is to look at the two accumulation functions, the simple discount accumulation function $a^{\text{s.d.}}(t) = \frac{1}{1-.08t}$ and the simple interest accumulation function $a^{\text{s.i}}(t) = 1 + .12t$ and determine $T > 0$ so that $a^{\text{s.d.}}(T) = a^{\text{s.i.}}(T)$. That is, we need to solve $\frac{1}{1-.08T} = 1 + .12T$. This equation is equivalent to the equation $-.0096T^2 + .04T + 1 = 1$ which has positive solution $T = \frac{.04}{.0096} = \frac{25}{6} \approx 4.16667$. From the investor's perspective, the simple discount investment account looks more and more attractive as time passes while the simple interest account becomes less and less desirable. Thus, the simple discount is preferable if money is kept on deposit for longer than $T \approx 4.16667$.

(5) (a) At the end of three years, the invested \$300 has grown in the simple interest fund to the accumulated amount $\$300[1+(.06)(3)] = \$354$. After an *additional* T years, during which time the money grows in the 8% simple discount account, the balance is is $\$354(1 - 08T)^{-1}$; this is because we are assuming that the simple discount account has just been opened. Thus, we are trying to determine $T+3$ where $\$354(1 - 08T)^{-1} = \$650$. This equation has solution $T = \frac{1}{.08}\left(1 - \frac{354}{650}\right) \approx 5.692307692$ and hence our answer is $T + 3 \approx 8.692307692 \approx 8.69231$ years.

(b) We seek i so that $\$300(1+i)^{T+3} = \$650$ where T is as in part (a). Note that $i = \left(\frac{650}{300}\right)^{\frac{1}{T+3}} - 1 \approx .0930271503 \approx 9.30272\%$.

(1.9) Compound discount

(1) The balance at the end of five years is $\$1{,}000(1 - .064)^{-5} \approx \$1,391.9407773 \approx \$1,391.94$.

(3) To bring back money one quarter, you multiply by$(1.068)^{-\frac{1}{4}}$. Therefore, the effective quarterly discount rate is $1 - (1.068)^{-\frac{1}{4}} \approx .016312423 \approx 1.63124\%$.

(5) We are given that in three years, $\$2{,}120 - \$250 = \$1{,}870$ grows to \$2,120. Therefore, $(1+i)^3 = \frac{2{,}120}{1{,}870}$. So, the interest for two years on \$380 is

$$\$380(1+i)^2 - \$380 = \$380\left[\left(\frac{2{,}120}{1{,}870}\right)^{\frac{2}{3}} - 1\right] \approx \$33.15508238 \approx \$33.16.$$

(7) We are given

$$\$320 = \$X[(1+i)^2 - 1] = \$X(i^2 + 2i) = \$Xi(i+2)$$

and

$$\$148 = \$Xd = \$X\left(\frac{i}{1+i}\right).$$

Rewrite this second equation as $\$Xi = \$148(1+i)$, and then substitute this new expression for $\$Xi$ into the first equation, thereby obtaining $\$320 = \$148(1+i)(i+2)$. It follows that $320 = 148i^2 + 444i + 296$, and hence $148i^2 + 444i - 24 = 0$. The quadratic formula then tells us that

$$i = \frac{-444 \pm \sqrt{(-444)^2 - 4(148)(-24)}}{2 \times 148}.$$

So that i is positive, we must take

$$i = \frac{-444 + \sqrt{(-444)^2 - 4(148)(-24)}}{2 \times 148} \approx .053113699 \approx 5.31137\%.$$

Consequently,

$$X = \frac{320}{i(i+2)} \approx \$2{,}934.475103 \approx \$2{,}934.48.$$

(1.10) Nominal rates of interest and discount

(1) The equivalent rates may be found as follows;

$1 - d = \left(1 - \frac{d^{(4)}}{4}\right)^4$ so $d = 1 - (1 - \frac{.08}{4})^4 \approx .07763184 \approx 7.76318\%.$

$\left(1 - \frac{d^{(3)}}{3}\right)^3 = \left(1 - \frac{d^{(4)}}{4}\right)^4$ so $d^{(3)} = 3\left[1 - (1 - \frac{.08}{4})^{\frac{4}{3}}\right] \approx .079732138 \approx 7.97321\%$

$1 + i = \left(1 - \frac{d^{(4)}}{4}\right)^{-4}$ so $i = (1 - \frac{.08}{4})^{-4} - 1 \approx .084165785 \approx 8.41658\%.$

$\left(1 + \frac{i^{(6)}}{6}\right)^6 = \left(1 - \frac{d^{(4)}}{4}\right)^{-4}$ so $i^{(6)} = 6\left[(1 - \frac{.08}{4})^{-\frac{4}{6}} - 1\right] \approx .08135748 \approx 8.13575\%.$

(3) Note that $i^{(12)} = 12 \times .5\% = 6\%$. Moreover, $i = 1 + .005)^{12} - 1 \approx .061677812 \approx 6.16778\%$. Also, $d = 1 - (1+i)^{-1} = 1 - 1 + .005)^{-12} \approx .05809466 \approx 5.80947\%$.

(5) Since $1 = \left(1 + \frac{i^{(m)}}{m}\right)\left(1 - \frac{d^{(m)}}{m}\right)$, $\frac{i^{(m)}}{m} - \frac{d^{(m)}}{m} = \frac{i^{(m)}}{m}\frac{d^{(m)}}{m}$. Multiplying this equation by by m^2 and substituting in the given values for $i^{(m)}$ and $d^{(m)}$, we find

$$m(.0469936613 - .046773854) = (.0469936613)(.046773854).$$

So,

$$m = \frac{(.046993661309)(.046773854)}{.0469936613 - .046773854} \approx 10.0000211 \approx 10.00002.$$

If we insist that m is an integer, then $m = 10$.

(7) (a) Every m years, money grows by a factor $(1+i)^m$ so $1+mi^{(\frac{1}{m})}=(1+i)^m$ and

$$i^{(\frac{1}{m})}=\frac{1}{m}\left[(1+i)^m-1\right].$$

(b) We are given a nominal interest rate of 6% for each year-and-a-half period. So, $(1+i)^{(\frac{3}{2})}=1+\frac{.06}{2/3}=1.09$. Therefore, $i=(1.09)^{\frac{2}{3}}-1\approx .059134217\approx 5.91342\%$.

(c) The discount factor for an m-year period may be expressed as $\left(1+mi^{(\frac{1}{m})}\right)^{-1}$ or as $1-md^{(\frac{1}{m})}$. These must be equal, so

$$d^{(\frac{1}{m})}=\frac{1}{m}\left[1-\left(1+mi^{(\frac{1}{m})}\right)^{-1}\right].$$

The discount factor for m years is also given by the expression $(1-d)^m$, and therefore $1-md^{(\frac{1}{m})}=(1-d)^m$. It follows that

$$d^{(\frac{1}{m})}=\frac{1}{m}\left[1-(1-d)^m\right].$$

(1.11) A friendly competition (Constant force of interest)

(1) We have $\delta=\ln(1+i)=\ln\left[\left(1-\frac{d^{(4)}}{4}\right)^{-4}\right]=-4\ln\left(1-\frac{d^{(4)}}{4}\right)=-4\ln\left(1-\frac{.032}{4}\right)\approx .032128687\approx 3.21287\%$.

(3) We compare the accounts by determining the annual effective interest rates i_A, i_B, and i_C of the three accounts; as the investor, you should choose the account with the highest rate. We calculate $i_B=(1.0044)^{12}-1\approx .054096687\approx 5.40967\%$, and $i_C=e^{.0516}-1\approx .052954476\approx 5.29545\%$. Since we are given $i_A=5.2\%$, it would provide the lowest accumulation while you should choose B.

(1.12) Force of interest

(1) (a) We begin by finding the accumulation function: $a(t)=e^{\int_0^t \delta_r\,dr}=e^{\int_0^t .05+.006r\,dr}=e^{.05t+.003t^2}$. So, the accumulated value of \$300 deposited at time 0 is $\$300a(3)=e^{.15+.027}=\$300e^{.177}\approx \$358.0893279\approx \$358.09$.

(b) If the \$300 deposit is made at time 4, then the deposit three years later is $\$300\frac{a(7)}{a(4)}=\$300e^{\int_4^7 .05+.006r\,dr}=\$300e^{(.05r+.003r^2)|_4^7}=\$300e^{.249}\approx \$384.8226099\approx \$384.82$.

(3) Note that $a(t)=e^{\int_0^t \delta_r\,dr}=e^{\int_0^t \frac{r^2}{1+r^3}dr}=e^{\frac{1}{3}(\ln 1+r^3)|_0^t}=e^{\ln\left[(1+t^3)^{\frac{1}{3}}\right]}=(1+t^3)^{\frac{1}{3}}$. Therefore, $a(4)=(65)^{\frac{1}{3}}$ and, the present vale of \$300 to be paid at time 4 is $\$300/(65)^{\frac{1}{3}}\approx \$74.6133952\approx \$74.61$.

(5) We find $\delta_t=\frac{d}{dt}a(t)=\frac{d}{dt}(t\ln(1+.02))+\ln(1+.03t)\frac{d}{dt}-\ln(1-.05t)\frac{d}{dt}=\ln(1.02)+\frac{.03}{1+.03t}-\frac{-.05}{1-.05t}$. Therefore $\delta_3=\ln(1.02)+\frac{.03}{1.09}+\frac{.05}{.85}\approx .106149092\approx 10.61491\%$.

(7) The 10% simple interest account has force of interest function $\delta_t=\frac{.1}{1+.1t}$, which is a decreasing function of T, and the 7% compound interest account has constant force of interest $\delta=\ln(1.07)$. If our goal is to maximize the accumulation at the end of five years or to maximize it at the end of ten years, we wish to always have our money in the account that has a higher force of interest. So, we should move our money when $\frac{.1}{1+.1t}=\ln(1.07)$; that is, at time $t=10\left(\frac{.1}{\ln 1.07)}-1\right)\approx 4.780076495\approx 4.78008$.

(9) We first need to determine the amount functions $A^{(A)}(t)$ and $A^{(B)}(t)$. Note that we have accumulation function

$$a^{(A)}(t) = e^{\int_0^t \delta_r{}^{(A)}\,dr} = e^{\int_0^t \frac{.08}{1+.08r}\,dr} = e^{\ln(1+.08t)} = 1 + .08t,$$

so $A^{(A)}(t) = 600(1 + .08t)$. Also,

$$a^{(B)}(t) = e^{\int_0^t \delta_r{}^{(B)}\,dr} = e^{\int_0^t .01r\,dr} = e^{.005t^2},$$

and thus $A^{(B)}(t) = 300e^{.005t^2}$. Therefore,

$$A^{(C)}(t) = A^{(A)}(t) + 2A^{(B)}(t) = 600(1 + .08t) + 600e^{.005t^2} = 600 + 48t + 600e^{.005t^2},$$

and

$$\delta_t^{(C)} = \frac{d}{dt}A^{(C)}(t)/A^{(C)}(t) = (48 + 6te^{.005t^2})/(600(1 + .08t) + 600e^{.005t^2}).$$

Evaluating at $t = 4$, we obtain $\delta_4^{(C)} = (48 + 24e^{.08})/(792 + 600e^{.08}) \approx .051317832 \approx 5.13178\%$.

(1.13) Note for those who skipped Section (1.11) and (1.12)

(1) Money grows each year by a factor of $e^{.0375}$; so the annual effective rate of interest is $e^{.0375} - 1 \approx .038211997 \approx 3.82120\%$.

(1.14) Inflation

(1) (a) Buying power grows by a factor of $\frac{1.042}{1.03} \approx 1.011650485$. Therefore, the real rate of interest is $\approx 1.16505\%$.

(b) In this case, buying power grows by a factor of $\frac{1.042}{1.046} \approx .99617590 = 1 + (-.003824092)$. So, the real rate of interest is approximately $-.38241\%$.

(3) Buying power grows by a factor of 1.0124 each year while, thanks to interest, money available for purchasing grows by $\left(1 - \frac{.03}{4}\right)^{-4}$. Thus, the real rate of inflation r satisfies the equation $1.0124 = \left(1 - \frac{.03}{4}\right)^{-4}/(1 + r)$. Equivalently, $r = \left(1 - \frac{.03}{4}\right)^{-4}(1.0124)^{-1} - 1 \approx .017948488 \approx 1.79485\%$.

(5) During the three-year period, purchasing power changes by a factor $\frac{\left(1+\frac{.024}{12}\right)^{36}}{(1.015)(1.028)(1.034)} \approx .995997571 = 1 - .004002429$. So, purchasing power falls by about .40024%.

Chapter 1 review problems

(1) The accumulated value is $\$6{,}208\left(1 - \frac{.023}{4}\right)^{-4\times 2}\left(1 + \frac{.03}{12}\right)^{12}(1 - .042)^{-3}e^{.046\times 2} \approx \$8{,}353.299474 \approx \$8{,}353.30$.

(3) The original annual effective interest rate i_0 correspond to force of interest δ_0 where $1 + i_0 = e^{\delta_0}$. Moreover, $\$1{,}039.98 = \$K(1 + i_0)^{-2} = \$Ke^{-2\delta_0}$. On the other hand, we are told that $\$1{,}060.78 = \$Ke^{-2\left(\frac{\delta_0}{2}\right)}$. So, $\$K(1 + i_0)^{-1} = \$1{,}060.78$. Therefore, $1 + i_0 = \frac{\$K(1+i_0)^{-1}}{\$K(1+i_0)^{-2}} = \frac{1{,}060.78}{1{,}039.98}$, and $K = (1{,}060.78)(1 + i_0) = \frac{(1{,}060.78)^2}{1{,}039.98} \approx 1{,}081.996008 \approx 1{,}082$. The interest rate i_0 is equivalent to an annual effective discount rate $d_0 = 1 - \frac{1}{1+i_0} = 1 - \frac{1{,}039.98}{1{,}060.78} = \frac{20.8}{1{,}060.78}$.

If we have a new annual effective discount rate d with $d = \frac{d_0}{2} = \frac{10.4}{1{,}060.78}$, then the present value of $\$K$ is $K(1 - d)^2 = \$1{,}082\left(1 - \frac{10.4}{1{,}060.78}\right)^2 \approx \$1{,}060.89$.

(5) The December 1, 2003 value of the first option is $\$6{,}000 + \frac{\$4{,}000}{1.05}$ while the value of the second option on that date is $\$12{,}000(1.05)^{-\frac{N}{12}}$. Setting these two values equal and then dividing by $12,000, we find $(1.05)^{-\frac{N}{12}} = .5 + \frac{1}{3(1.05)} \approx .817460317$. Taking natural logaithms, we obtain $-\frac{N}{12}\ln(1.05) \approx \ln(.817460317)$. So, $N \approx -12\ln(.817460317)/\ln(1.05) \approx 49.5721846 \approx 49.57218$.

(7) (a) The accumulation function satisfies $a(0) = 1$, so $c = 1$. To determine the values of the constants a and b, we consider the conditions $i_3 = 50/1{,}088$ and $d_4 = 54/1{,}192$ simultaneously. Note that

$$\frac{50}{1{,}088} = i_3 = \frac{a(3) - a(2)}{a(2)} = \frac{(9a + 3b + 1) - (4a + 2b + 1)}{4a + 2b + 1} = \frac{5a + b}{4a + 2b + 1}.$$

So,

$$200a + 100b + 50 = 50(4a + 2b + 1) = 1{,}088(5a + b) = 5,440a + 1{,}088b.$$

Equivalently, $a = \frac{50 - 988b}{5{,}240}$. On the other hand,

$$\frac{54}{1{,}192} = d_4 = \frac{a(4) - a(3)}{a(4)} = \frac{(16a + 4b + 1) - (9a + 3b = 1)}{16a + 4b + 1} = \frac{7a + b}{16a + 4b + 1}.$$

Therefore,

$$864a + 216b + 54 = 54(16a + 4b + 1) = 1{,}192(7a + b) = 8,344a + 1{,}192b.$$

It follows that $54 = 7480a + 976b = 7480\left(\frac{50-988b}{5{,}240}\right) + 976b$. Thus,

$$282{,}960 = 54(5240) = 7480(50 - 988b) + 976b(5,240) = 374{,}000 - 7{,}390{,}240b + 5{,}114{,}240b,$$

and

$$b = \frac{374{,}000 - 282{,}960}{7{,}390{,}240 - 5{,}114{,}240} = \frac{91{,}040}{2{,}276{,}000} = .04.$$

Moreover, $a = \frac{50-988b}{5{,}240} = \frac{50-988(.04)}{5{,}240} = .002$.

(b) The value at $t = 3$ of $1,000 to be paid at $t = 8$ is given by the expression $\left(\frac{\$1{,}000}{a(8)}\right)a(3)$. Since

$$a(3) = 9a + 3t + 1 = 9(.002) + 3(.04) + 1 = 1.138$$

and

$$a(8) = (64a + 8b + 1)(1 + .05(8 - 6)) = [64(.002) + 8(.04) + 1](1.1) = 1.5928,$$

this is equal to $\$714.4650929 \approx \$714.47$.

(9) Note that

$$f(n) = i_{n+i} + 1 = \frac{a(n+1) - a(n)}{a(n)} + \frac{a(n)}{a(n)} = \frac{a(n+1)}{a(n)}.$$

Since

$$a(t) = e^{\int_0^t \delta_r\,dr} \quad \text{for} \quad 2 \le n \le 7,$$

we find

$$\begin{aligned} f(n) &= e^{\int_0^{n+1} \delta_r\,dr} / e^{\int_0^{n} \delta_r\,dr} \\ &= e^{\int_n^{n+1} \delta_r\,dr} = e^{\int_n^{n+1} \frac{4}{r-1}\,dr} \\ &= e^{4\ln(r-1)|_n^{n+1}} = e^{\ln(n^4) - \ln[(n-1)^4]} = e^{\ln\left(\frac{n^4}{(n-1)^4}\right)} = \left(\frac{n}{n-1}\right)^4. \end{aligned}$$

C H A P T E R 2

Equations of value and yield rates

(2.2) Equations of value for investments involving a single deposit made under compound interest

(1) The accumulation function is $a(t) = (1 - .04)^{-t}$, and a time 3 equation of value is $\$K(1 - .04)^{-3} = \$982$. Therefore, $K = \$982(.96)^3 \approx \$868.81$.

(3) Let T denote the time, measured in years, that has elapsed since the deposit. With time again measured in years, we have accumulation function $a(t) = (1 + \frac{.032}{12})^{12t}$. Therefore, $\$1{,}965.35 = \$1{,}800(1 + \frac{.032}{12})^{12T}$. It follows that

$$T = \frac{1}{12}\left[\ln\left(\frac{1,965.35}{1,800}\right)\Big/\ln\left(1 + \frac{.032}{12}\right)\right] \approx 2.750025245 \approx 2.75003.$$

(5) We seek an approximation to T where $3 = (1 + i)^T$. An exact solution is given by

$$T = \frac{\ln 3}{\ln(1+i)} = \left(\frac{\ln 3}{\ln(1+i)}\right)\left(\frac{i}{i}\right) = \ln 3 f(i)/i \quad \text{where} \quad f(i) = \frac{i}{\ln(1+i)}.$$

As observed in Example (2.2.4), the function $f(i)$ grows very slowly for positive i close to zero. Therefore,

$$T \approx (\ln 3)f(.08)/i = \frac{(\ln 3)(.08)}{\ln(1.08)}/i \approx \frac{1.141993167}{i} \approx \frac{1.14}{i}.$$

We therefore suggest a "rule of 114". With this rule, you estimate a period of $\frac{114}{4} = 28.5$ years for money to triple at 4% and a period of $\frac{114}{12} = 9.5$ years for money to triple at 12%. In fact, the times for tripling are $\frac{\ln 3}{\ln(1.04)} \approx 28.01102276$ and $\frac{\ln 3}{\ln(1.12)} \approx 9.694035413 \approx 9.69404$.

(2.3) Equations of value for investments with multiple contributions

(1) Denote the time of the loan by $t = 0$. From Sidney's perspective, the financial arrangement may be represented by the following timeline.

CASH FLOW:	\$12,000	−\$4,000	$-X$	−\$3,000
TIME:	0	1	2	3

Valuing each payment at time 2, we have the time 2 equation of value

$$\$12{,}000(1+i)^2 = \$4{,}000(1+i) + X + \$3{,}000(1+i)^{-1}.$$

But, $(.94)^{-1} = (1-d)^{-1} = 1 + i$. Therefore, $\$12{,}000(.94)^{-2} = \$4{,}000(.94)^{-1} + X + \$3{,}000(.94)$. It follows that

$$X = \$12{,}000(.94)^{-2} - \$4{,}000(.94)^{-1} - \$3{,}000(.94) \approx \$6{,}505.486646 \approx \$6{,}505.49.$$

(3) Denoting the time of the loan by $t = 0$, the cash flows may be depicted on a timeline (*from Esteban's perspective*) as follows;

CASH FLOW :	\$20,000	−\$10,000	−\$12,000
TIME:	0	T	$2T$

We therefore have a time $2T$ equation of value

$$\$20{,}000(1.06)^{2T} = \$10{,}000(1.06)^{T} + \$12{,}000.$$

Letting $x = (1.06)^T$, we may rewrite this as $20x^2 - 10x - 12$. The quadratic formula then yields $x = \frac{10 \pm \sqrt{(-10)^2 - 4(20)(-12)}}{2(20)} = \frac{10 \pm \sqrt{1{,}060}}{40}$. Since, $x = (1.06)^T$ is positive, we must have $x = \frac{10+\sqrt{1{,}060}}{40} \approx 1.06394103$. It follows that $T = \frac{\ln x}{\ln 1.06} \approx 1.063688479 \approx 1.06369$ years.

(5) Denoting the present time as time $t = 0$, the scheduled loan repayments are pictured below.

	\$8,000	\$9,000		\$20,000
0	1	2	3	4

According to the method of equated time, $\overline{T}$ is a dollar-weighted average of the payment times; that is, with time measured in years,

$$\overline{T} = \left(\frac{\$8{,}000}{\$37{,}000}\right)1 + \left(\frac{\$9{,}000}{\$37{,}000}\right)2 + \left(\frac{\$20{,}000}{\$37{,}000}\right)4 = \left(\frac{106{,}000}{37{,}000}\right) \approx 2.864864865 \approx 2.86487.$$

On the other hand, the exact time T for repayment so that the present value of the renegotiated payment matches that of the original payments must satisfy the time 0 equation of value

$$\$8{,}000(1.05)^{-1} + \$9{,}000(1.05)^{-2} + \$20{,}000(1.05)^{-4} = \$37{,}000(1.05)^{-T}.$$

Therefore,

$$T = -\ln\left(\frac{8{,}000(1.05)^{-1} + 9{,}000(1.05)^{-2} + 20{,}000(1.05)^{-4}}{37{,}000}\right)\Big/\ln(1.05) \approx 2.824807661. \approx 2.82481$$

(7) (a) We desire the point-slope form of the tangent line to the curve $y = \ln x$ at $(A, \ln A)$. More generally, the point slope-form of the tangent line to the curve $y = f(x)$ at the point (x_0, y_0) is $y - y_0 = f'(x_0)(x - x_0)$. Since $\frac{d}{dx}\ln x\,|_{x=A} = \frac{1}{x}\,|_{x=A} = \frac{1}{A}$, the tangent line is $y - \ln A = \frac{1}{A}(x - A)$.

(b) Note that if $x > 0$ so that $\ln x$ is defined, then the second derivative $\frac{d^2}{dx^2}\ln x$ satisfies

$$\frac{d^2}{dx^2}\ln x = \frac{d}{dx}\frac{1}{x} = \frac{d}{dx}x^{-1} = -x^{-2} < 0.$$

This means that the function $y = \ln x$ is concave down; that is, the curve lies below it's tangent line except at the point of intersection of the tangent line and the curve. Recalling the equation of the tangent line that we found in part (a), we thus have

$$y = \ln x \leq \frac{1}{A}(x - A) + \ln A;$$

equality holds only at $x = A$.

(c) Since $G = (\prod_{k=1}^{n} b_k)^{\frac{1}{n}}$, we have $\ln G = \ln\left((\prod_{k=1}^{n} b_k)^{\frac{1}{n}}\right) = \frac{1}{n}\ln(\prod_{k=1}^{n} b_k) = \frac{1}{n}(\sum_{k=1}^{n}\ln b_k)$. Therefore, using the inequality established in part (b), $\ln G \leq \frac{1}{n}(\sum_{k=1}^{n}\frac{1}{A}(b_k - A) + \ln A)$, and equality holds if and only if $b_1 = b_2 = \cdots = b_n = A$. But

$$\frac{1}{n}\left(\sum_{k=1}^{n}\frac{1}{A}(b_k - A) + \ln A\right) = \frac{1}{n}\left(\sum_{k=1}^{n}\frac{1}{A}(b_k - A)\right) + \frac{1}{n}\left(\sum_{k=1}^{n}\ln A\right) = \frac{1}{n}\left(\sum_{k=1}^{n}\frac{1}{A}(b_k - A)\right) + \ln A,$$

so we have $\ln G \leq \frac{1}{n}\sum_{k=1}^{n}A^{-1}(b_k - A) + \ln A$ with strict inequality unless all the b_k's equal A.

(d) In order to prove $\frac{1}{n}\sum_{k=1}^{n}A^{-1}(b_k - A) + \ln A = \ln A$, it suffices to show $\sum_{k=1}^{n}A^{-1}(b_k - A)A = 0$. But since A is constant , $\sum_{k=1}^{n}A^{-1}(b_k - A)A = \frac{\sum_{k=1}^{n}b_k}{A} - \frac{nA}{A} = \frac{nA}{A} - \frac{nA}{A} = 0$; here the equality $\sum_{k=1}^{n}b_k = A$ follows from the definition of A as the average of the n b_k's.

(e) Combining the results established in parts (c) and (d), we obtain $\ln G \leq \ln A$ and equality holds if and only if each b_K takes on the value A. Since the natural logarithm is an increasing function, this equality is equivalent with the inequality $G \leq A$ and we have equality exactly if all the b_k's are equal (*to their average*).

(9) We first look for T satisfying the time 0 equation of value $\$1{,}000 - \$4{,}000(1.01)^{-1} + \$2{,}000(1.01)^{-2} = -\$1{,}000(1.01)^{-T}$. The left hand side of this equation is approximately -\$999.8039408, and therefore $T = -\frac{\ln(.9998039408)}{\ln(1.01)} \approx .01970572 \approx .01971$. On the other hand, $\overline{T} = -1(0) + 4(1) + -2(2) = 0$, so $T > \overline{T}$.

(11) (a) Looking at a time 0 equation of value, we seek the amount X so that $\$400 + \frac{\$300}{1+4(.06)} = \frac{X}{1+8(.06)}$. Therefore, $X = 1.48\left(\$400 + \frac{\$300}{1+4(.06)}\right) \approx \$950.0645161 \approx \$950.06$.

(b) This time we look for Y satisfying the time 8 equation of value $\$400[1+.06(8)] + \$300[1+.06(4)] = Y$, namely $Y = \$964$.

(c) In part (a), we do not have a single accumulation function governing the growth of all money. Instead, the choice of accumulation function depends on the time of deposit. Since simple interest has a decreasing force of interest as time goes by, newly deposited money earns interest more quickly than money that has been on deposit for a while.

(d) The condition that you have compound interest is equivalent to the requirment that the force of interest is constant. Thereofre, the answers to parts (a) and (b) will be equal. if we look at a time 8 equation of value as in (b), we find $\$400(1.06)^8 + \$300(1.06)^4 = Y.$, and $Y \approx \$1{,}016.282318 \approx \$1{,}016.28$.

(2.4) Investment return

(1) The yield rate I must satisfy the time 8 equation of value $\$3{,}000(1+I)^8 + \$8{,}000 = \$10{,}000(1+I)^4$. Letting $x = (1+I)^4$, this equation is equivalent to the quadratic equation $3x^2 - 10x + 8 = 0$ which, according to the quadratic formula has the two soluitons

$$x = \frac{10 + \sqrt{(-10)^2 - 4(3)(8)}}{6} = \frac{10+2}{6} = 2 \quad \text{and} \quad x = \frac{10 - \sqrt{(-10)^2 - 4(3)(8)}}{6} = \frac{10-2}{6} = \frac{4}{3}.$$

It follows that

$$I = x^{\frac{1}{4}} = 2^{\frac{1}{4}} - 1 \approx .189207115 \quad \text{or} \quad I = x^{\frac{1}{4}} = \left(\frac{4}{3}\right)^{\frac{1}{4}} - 1 \approx .07456932.$$

If we denote the two possible values for the yield rate I by i and j as was done in the statement of the problem, it follows that $|i - j| \approx .189207115 - .07456932 \approx .114637183 \approx 11.46372\%$.

(3) The yield rate i satisfies the time 4 equation of value $\$150{,}000(1+i)^4 + \$40{,}000 = \$210,000(1+i)^2$. Letting $x = (1+i)^2$, this equation is equivalent to $15x^2 - 21x + 4 = 0$, and the quadratic equation gives

$$x = \frac{21 \pm \sqrt{(-21)^2 - 4(15)(4)}}{30} = \frac{21 \pm \sqrt{201}}{30}.$$

But x is the square of a real number, hence must be positive. Thus $x = \frac{21+\sqrt{201}}{30} \approx 1.172581563$, and $i = \sqrt{x} - 1 \approx .082858053 \approx 8.28581\%$.

If you have a BA II Plus calculator, you may also find this using the **Cash Flow worksheet**. Just key [2ND] [CF] [2ND] [CLR WORK] [1] [5] [0] [0] [0] [0] [ENTER] [↓] [2] [1] [0] [0] [0] [0] [+/−] [ENTER] [↓] [↓] [4] [0] [0] [0] [0] [ENTER] [IRR] [CPT], followed by [%] [+] [1] [=] [yˣ] [√x] [−] [1] [=]; the first string of keys computes the effective two-year yield rate expressed as a percent, and the next sequence convert to the one year yield rate in decimal notation.

(5) A time zero equation of value for the given financial arrangement is

$$\$8{,}572.80 + \$28{,}500v^2 = \$27{,}074v + \$10{,}000.$$

If $v = .94$, the two sides of the equation have the common value \$33,755.40; if $v = .95$, the two sides of the equation have the common value \$34,294.05; if $v = .96$, the two sides of the equation have the common value \$34,838.40. Moreover, since $i = \frac{1}{v} - 1$, the corresponding yield rates are $\frac{1}{.94} - 1 \approx .063829787 \approx 6.38298\%$, $\frac{1}{.95} - 1 \approx .052631579 \approx 5.26316\%$, and $\frac{1}{.96} - 1 \approx .041666667 \approx 4.16667\%$ respectively.

(7) We proceed as in the proof of Important Fact (2.4.9). Once again, interchanging the roles of borrower and lender if necessary, we may assume that $B_{t_1}(i) = C_{t_1}$ is positive. Then, the inequality $-1 < j \leq i$ forces $C_{t_1}(1+i) \leq C_{t_1}(1+j)$, and equality holds only when $i = j$. It follows that $B_{t_2}(i) = B_{t_1}(i)(1+i) - C_{t_2} \leq B_{t_1}(i)(1+j) - C_{t_2} = B_{t_1}(i)(1+j) - C_{t_2} = B_{t_2}(j)$, equality once again requiring $i = j$. Continuing in this manner, using an inductive argument if you wish to be formal, we find $B_{t_k}(i) \leq B_{t_k}(j)$ for $k = 2, 3, \ldots, n$, equality holding only if the rates i and j are equal. But, we are given that $B_{t_n}(i) = B_{t_n}(j)$, so $i = j$.

(9) First open and clear the **Cash Flow worksheet** by keying [2ND] [CF] [2ND] [CLR WORK]. Next, key [2] [0] [0] [0] [0] [+/−] to obtain "CFo = −20,000". Now enter the actual amounts of payments 1 though 23 in registers C01 through C23. The k-th payment is for $\$375 + \$25k$, so the 24th payment is for \$975, the 33rd is for \$1,200, and the sum of the 24th through 33rd payments (which are in arithmetic progression), is $\frac{10}{2}(\$975 + \$1{,}200) = \$10{,}875$. If you have a BA II Plus calculator, then the entry in the C24 register should be \$10,875. If you have a BA II Plus Professional which has room for the 24th through 31st entries individually, you may enter them in registers C24 through C31 and then put $\$1{,}175 + \$1{,}200 = \$2{,}375$ in C32. Each frequency register should hold the entry 1. Now key [IRR] [CPT]. Using the BA II Plus calculator's twenty-four registers as suggested, you should obtain "IRR = 1.600129514", while if you used thirty-two registers, "IRR = 1.457288206" results. In each case, some of the money came in early, so the actual monthly yield rate is a bit lower; however, it must be very close to 1.45. In fact, at 1.45, the net

present value is about 15.9, at 1.452% it is about 8.56, at 1.4.54% it is about 1.21, at 1.4545% it is about −.62, at 1.45435% it is about −.007, and at 1.45436%, it is roughly .106. So, the monthly yield rate is between 1.45435% and 1.45436%, and the annual yield is between 18.9181305% and 18.91827118%. So, it is about 18.91813% or, to the nearest hundredth of a percent, 18.92%.

(2.5) Reinvestment considerations

(1) Kathy receives \$8,000 at $t = 0$, then repays \$6,000 at $t = \frac{3}{2}$ and \$4,000 at $t = 3$. A time 3 equation of value for Kathy's three-year loan is

$$\$8{,}000(1+i_K)^3 - \$6{,}000(1+i_K)^{\frac{3}{2}} - \$4{,}000 = 0.$$

So, $4(1+i_K)^3 - 3(1+i_K)^{\frac{3}{2}} - 2 = 0$. This is quadratic in $(1+i_K)^{\frac{3}{2}}$, and the quadratic formula gives

$$(1+i_K)^{\frac{3}{2}} = \frac{3 \pm \sqrt{3^3 - 4(4)(-2)}}{2(4)}.$$

Since $(1+i_K)^{\frac{3}{2}} > 0$, we have $(1+i_K)^{\frac{3}{2}} \approx 1.17539053$ and $i_K \approx 11.37510\%$. If you have a BA II Plus calculator, you may obtain the one-and-a-half year rate 17.539053% more easily by using the **Cash Flow worksheet**. Use CF0 = 8,000, C01= −6,000, F01 = 1, C02 = −4,000, and F02 = 1.

Angela invests \$8,000. Due to reinvestment, she does not receive any money back until $t = 3$, at which time she gets $\$4{,}000 + \$6{,}000(1.06)^{\frac{3}{2}} \approx \$10{,}548.02$. Therefore, Angela's yield rate i_A satisfies $\$8{,}000(1+i_A)^3 = \$10{,}548.02$, and $i_A \approx 9.65463\%$.

(3) Enterprise A loans a total of \$23,000 (\$17,000 + \$6,000) on January 15, 2000. The \$7,000 that comes in two years later is reinvested for two years at 5%, producing $\$7{,}000(1.05)^2 = \$7{,}717.50$. So, at the end of four years, Enterprise A receives \$22,500 + \$7,717.50 = \$30,217.50. Denoting its annual yield rate by i_A, we have $\$23{,}000(1+i_A)^4 = \$30{,}217.50$. It follows that $i_A \approx 7.06134\%$.

Enterprise B receives \$6,000 and repays \$7,000 two years later. So, its annual effective interest rate i_B satisfies $\$6{,}000(1+i_B)^2 = \$7{,}000$, so $i_B \approx 8.01234\%$.

Enterprise C receives \$17,000 and repays \$22,500 four years later. So, its annual effective interest rate i_C satisfies $\$17{,}000(1+i_C)^4 = \$22{,}500$, so $i_B \approx 7.25891\%$.

(2.6) Approximate dollar-weighted yield rates

(1) (a) You have a choice of the time to value the cashflows. If you choose time zero, the equation of value is

$$\$10{,}832 + \$2{,}000(1+i)^{-\frac{3}{12}} - \$7{,}000(1+i)^{-\frac{5}{12}} = \$12{,}566(1+i)^{-2}.$$

(The problem does not ask you to compute the annual yield rate i, but you may wish to. Using the **Cash Flow worksheet** of the BA II Plus calculator, you may find the solution i to this equation; you will first find a monthly yield rate by entering CF0 = −10,832, C01 = 0, F01 = 2, CO2 = −2,000, F02 = 1, C03 = 0, F03 = 1, C04 = 7,000, F04 = 1, C05 = 0, F05 = 18, C06 = 12,566, F06=1, then keying **IRR** **CPT**. The resulting monthly yield rate 2.743970879% is equivalent to the annual yield rate 38.38090%. This rate may also be found by "guess and check" or by "Newton's method".)

(b) We have $A = \$10{,}832$, $C = \$2{,}000 - \$7{,}000 = -\$5{,}000$, and $B = \$12{,}566$. Therefore, $I = B - A - C = \$6{,}734$. Note that $C_{\frac{3}{24}} = \$2{,}000$, $C_{\frac{5}{24}} = -\$7{,}000$, and $C_t = 0$ for all other t. Therefore, Equation (2.6.5) gives

$$j \approx \frac{I}{A + \sum_{t\in(0,1)} C_t(1-t)} = \frac{\$6{,}734}{\$10{,}832 + \$2{,}000\left(\frac{21}{24}\right) - \$7{,}000\left(\frac{19}{24}\right)} \approx .956488803,$$

but this is a two-year rate. The corresponding dollar-weighted annual yield rate is $(1+j)^{\frac{1}{2}} \approx .39874539 \approx 39.87454\%$.

If we use Equation (2.6.8) to estimate the two-year dollar-weighted yield, we find

$$j \approx \frac{2I}{A+B-I} = \frac{2(\$6{,}734)}{\$10{,}832 + \$12{,}566 - \$6{,}734} \approx .808209318.$$

The corresponding approximate dollar-weighted annual yield is $(1+j)^{\frac{1}{2}} \approx .344696737 \approx 34.46967\%$.

(3) We are given an initial balance $A = \$290{,}000$ and a final balance $B = \$448{,}000$. Furthermore, the total interest is $I = \$34{,}000$. Thus, $C = B - A - I = \$448{,}000 - \$290{,}000 - \$34{,}000 = \$124{,}000$. The annual yield rate was given to be 5.4%, so the two-year yield rate is $(1.054)^2 - 1 = .110916$. Thus, according to Equation (2.6.6), the average date of contribution k satisfies

$$.110916 = \frac{\$34{,}000}{\$290{,}000 + \$124{,}000(1-k)}.$$

Therefore, $k \approx .866626764$. Observe that we are considering the two-year period from January 1, 1995 through the end of 1996. This consists of $365 + 366 = 731$ days. Since $731k \approx (731)(.866626764) \approx 634$, the average date of payment is September 26, 1996. (You might use the BA II Plus calculator **Date worksheet** to calculate this; use an actual day count with "DT1 = 1 - 01 - 1995" and "DBD = 634".)

(2.7) Fund performance

(1) Let j_{tw} denote the time-weighted yield for the four-year period from January 1, 1988 to January 1, 1992; this is a four-year rate. During this period, the balance grows as follows:

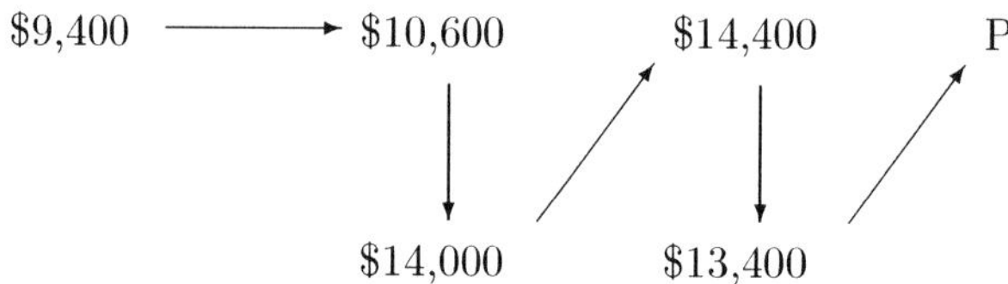

Here, the left-hand vertical arrow is due to a \$2,400 deposit, while the right-hand vertical arrow corresponds to a \$1,000 withdrawal. The vertical arrows are not reflected in the time-weighted yield rate j_{tw}, while the other arrows each give us a ratio. We have

$$j_{tw} = \frac{\$10{,}600}{\$9{,}400} \times \frac{\$14{,}400}{\$14{,}000} \times \frac{P}{\$13{,}400}.$$

We seek the equivalent annual rate, namely $i_{tw} = (1+j_{tw})^{\frac{1}{4}} - 1 = \left(\frac{15{,}264}{163{,}748{,}000} P\right)^{\frac{1}{4}} - 1.$

(3) Let X denote the end-of-year balance. During the year, the balance grows as follows:

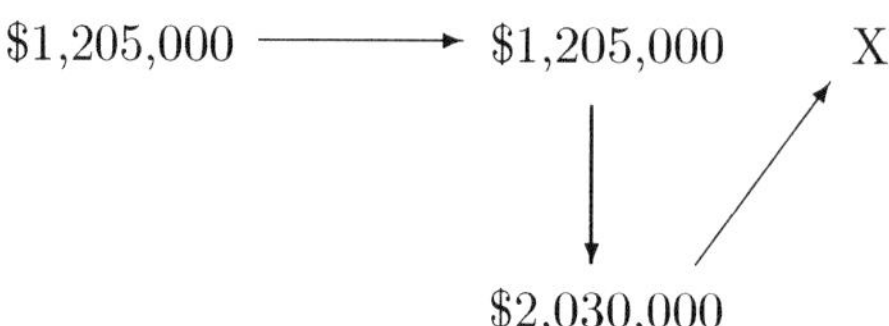

Here, the vertical arrow reflects a \$800,000 deposit, while the two other arrows reflect changes due to fund performance. So, the time-weighted yield rate for the one year period is 16%, so

$$1.16 = \frac{1{,}230{,}000}{\$1{,}205{,}000} \times \frac{X}{\$2{,}030{,}000}.$$

Therefore, $X \approx \$2{,}306{,}938.21$.

Chapter 2 review problems

(1) (a) A time 0 equation of value is

$$\$20{,}000 = \frac{\$4{,}000}{1+r} + \frac{\$18{,}000}{1+3r} = 0,$$

and $r \approx 3.82448\%$. This may be rewritten as

$$\$20{,}000 = \frac{\$4{,}000(1+3r) + \$18{,}000(1+r)}{(1+r)(1+3r)}$$

or as

$$\$20{,}000(1+4r+3r^2) = \$4{,}000 + \$12{,}000r + \$18{,}000 + \$18{,}000r.$$

So, $60{,}000r^2 + \$50{,}000r - 2{,}000 = 0$. Applying the quadratic formula to the equation $60r^2 + \$50r - 2 = 0$ and using the fact that $r > 0$, we obtain $r = \frac{-50+\sqrt{(-50)^2-4(60)(-2)}}{2(60)} \approx .038244802 \approx 3.82448\%$.

(b) A time 3 equation of value is

$$-\$20{,}000(1+i)^3 + \$4{,}000(1+i)^2 + \$18{,}000 = 0,$$

and a time 0 equation of value is

$$-\$20{,}000 + \$4{,}000(1+i)^{-1} + \$18{,}000(1+i)^{-3} = 0.$$

There is a unique yield rate since all deposits take place before any withdrawal. This may be estimated by using "guess and check" as dictated in the problem (or Newton's method); you need an initial guess, and you might reasonably start with a rate slightly below that found in part (a), say 3.75%. Let

$$f(i) = -20{,}000 + 4{,}000(1+i)^{-1} + 18{,}000(1+i)^{-3},$$

a decreasing function of i. Since $f(.0375) \approx -26.688$, and $f(.0365) \approx 23.727$, the desired rate is between 3.65% and 3.75%. Thus, the rate to the nearest hundredth of a percent is 3.70%.

(c) As in part (b), let

$$f(i) = -20{,}000 + 4{,}000(1+i)^{-1} + 18{,}000(1+i)^{-3}.$$

Then, $f(.0375) \approx -25.68825979$, $f'(i) = -4{,}000(1+i)^{-2} - 54{,}000(1+i)^{-4}$, and $f'(.0375) \approx -50{,}0509.30951$. Therefore, with an initial approximation of $i_1 = .0375$, our next approximation is $i_2 = .0375 - \frac{-25.68825979}{-50{,}0509.30951} \approx .036971617$. Next observe that $f(i_2) \approx -.072857695$ and $f'(i_2) \approx -50{,}420.86802$, so our next approximation is $i_3 = .036971617 - \frac{-.072857695}{-50{,}420.86802} \approx .036973062$. So, i_1 and i_3 are each about .03697, and hence the rate to the nearest hundredth of a percent is 3.70%.

(d) It is somewhat easier to find the root using the **Cash Flow worksheet** with CF0 = −20,000, C01 = 4,000, F01 = 1, C02 = 0, F02 = 1, C03 = 18,000, and F03 = 1; keying **IRR** **CPT** then gives "IRR = 3.697017203", so the rate to the nearest millionth of a percent is 3.697017%.

(3) Elyse's time 2 equation of value is

$$\$16{,}312(1+i)^2 = \left(\frac{1}{2}(\$8{,}000)\right)(1+i) + \left[\left(\frac{1}{2}(\$8{,}000)\right)(1.05) + \$10{,}000\right] = \$4{,}000(1+i) + \$14{,}200.$$

The quadratic formula (or **Cash Flow worksheet**) yields $i \approx .06364990969 \approx 6.36499\%$.

(5) (a) Let i_{tw} denote the desired annual rate, and let j_{tw} give the three-year time-weighted rate. Then $i_{tw} = (1+j_{tw})^{\frac{1}{3}} - 1$. The growth of the balance over the three-year period looks like

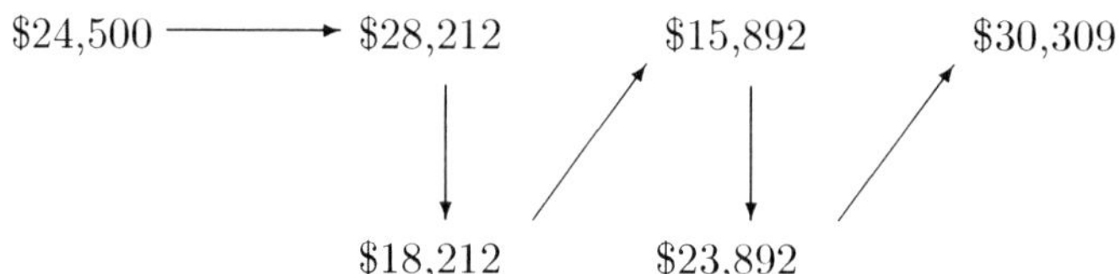

where the vertical arrows reflect a \$10,000 withdrawal and a \$8,000 deposit, respectively. The non-vertical arrows are used to determine the time-weighted yield; we have

$$1 + j_{tw} = \frac{\$28{,}212}{\$24{,}500} \times \frac{\$15{,}892}{\$18{,}212} \times \frac{\$30{,}309}{\$23{,}892} \approx 1.274699476,$$

so $i_{tw} = 1.274699476^{\frac{1}{3}} - 1 \approx 8.42664\%$.

(b) With notation as in Section (2.6), $A = \$24{,}500$, $B = \$30{,}309$, $C_{\frac{16}{24}} = -\$10{,}000$, $C_{\frac{23}{24}} = \$8{,}000$, $C = -\$10{,}000 + \$8{,}000 = -\$2{,}000$, and $I = B - A - C = \$7{,}809$. We thus have a three-year dollar wieighted yield rate

$$j = \frac{\$7{,}809}{\$24{,}500 - \$10{,}000\left(1 - \frac{16}{24}\right) + \$8{,}000\left(1 - \frac{16}{24}\right)} \approx .357664122\,.$$

The annual time-weighted yield is $(1+j)^{\frac{1}{3}} - 1 \approx 10.72970\%$.

(c) A time 0 equation of value for the Abiyote's three-year investment is

$$\$24{,}500 = \$10{,}000(1+i)^{-\frac{16}{24}} - \$8{,}000(1+i)^{-\frac{23}{24}} + \$30{,}309(1+i)^{-3},$$

so we seek a solution to this equation. Given our approximation in part (b), you might find an approximation to the solution, correct to the nearest millionth of a percent, by either "guess and check" or Newton's method, starting with an initial estimate of 10.7%; this is somewhat time consuming. It is quicker to use the **Cash Flow worksheet** of the BA II Plus calculator. You should enter CF0 = −24,500, C01 = 0, F01 = 15, C02 = 10,000, F02 = 1, C03 = 0, F03 = 6, C04 = −8,000, F04 = 1, C05 = 0, , F05 = 12, C06= 30,309, and then key **IRR** **CPT** to obtain a monthly yield rate of $j = .849572986\%$. The equivalent annual effective yield rate is $(1+j)^{12} - 1 \approx 10.68499865\% \approx 10.68500\%$.

(d) In part (a), we calculated the time-weighted yield rate for the Utopia Fund to be about 8.42664%, while in part (c) we found Abiyote's dollar yield rate to be 10.72970% which is higher. This is to be expected, because Abiyote made deposits just before the fund experienced upswings as well as a large withdrawal just before it took a downward turn.

CHAPTER 3

Annuities (annuities certain)

(3.2) Annuities-immediate

(1) There are 60 months in a five-year period. We are thus first asked to compute $\$3{,}000a_{\overline{60}|j}$, where $j = \frac{3\%}{12} = .0025$, and then to find $\$3{,}000a_{\overline{60}|q}$, where $q = \left[(1-.03)^{-\frac{1}{12}} - 1\right] \approx .00254149142$. We calculate

$$\$3{,}000a_{\overline{60}|j} = \frac{1-(1.0025)^{-60}}{.0025} \approx \$166{,}957.07,$$

and

$$\$3{,}000a_{\overline{60}|q} = \frac{1-(1+q)^{-60}}{q} = \frac{1-(.97)^{5}}{q} \approx \$166{,}751.66.$$

Note that $a_{\overline{60}|j} > a_{\overline{60}|q}$. This is because $q > j$ and therefore the payments are discounted more when calculating the present value $a_{\overline{60}|q}$ than when calculating $a_{\overline{60}|j}$.

It is efficient to calculate these symbols using the **TVM worksheet**. With your calculator in END mode and P/Y=C/Y=1, you may calculate $\$3,000a_{\overline{60}|j}$ by keying

[6] [0] [N] [•] [2] [5] [I/Y] [3] [0] [0] [0] [+/−] [PMT] [0] [FV] [CPT] [PV].

If you then change the entry in I/Y from j to q, once again entering the monthly interest rate as a percent, keying [CPT] [PV] will give you $\$3{,}000a_{\overline{60}|q}$.

(3) If there is a level annual effective interest rate of 5% throughout the eighteen years, we are looking for the least possible payment amount A so that $As_{\overline{18}|5\%} \geq \$150{,}000$. Since $s_{\overline{18}|5\%} \approx 28.13238467$, we want the smallest possible A with $A \geq \frac{\$150{,}000}{28.13238467} \approx \$5{,}331.933348$. Since the Browns make payments of amount A, this amount must be an integer number of cents. Thus, $A = \$5{,}331.94$. In fact, $\$5{,}331.94s_{\overline{18}|5\%} \approx \$150{,}000.1871 \approx \$150{,}000.19$, so the last payment may be reduced by 19 cents to \$5,331.75.

We next consider what happens if the annual effective rate of interest is level at 5% for the first ten years, then drops to 4.5%. Under this scenario, at time 10 the Browns have just made their tenth annual deposit of \$5,331.94, and these ten payments have accumulated to $\$5{,}331.94s_{\overline{10}|5\%}$. These will then accumulate for eight years at an annual effective interest rate of 4.5%, so as to reach $(1.045)^8\$5{,}331.94s_{\overline{10}|5\%} \approx 95{,}372.56372$. The Browns must thus make eight more deposits of amount B so that $Bs_{\overline{8}|5\%} \approx \$150{,}000 - 95{,}372.56372 \approx \$54{,}627.43628$. We calculate $B \approx \$5{,}823.812$. So as not to fall short, we round up to \$5,823.82, an increase of \$491.88. However, this will produce a final balance of $(1.045)^8\$5{,}331.94s_{\overline{10}|5\%} + \$5{,}823.812s_{\overline{8}|5\%} \approx \$150{,}000.07$, so the final payment only needs to be increased by $\$491.88 - .07 = \$491.81$.

(5) As usual, denote the annual effective interest rate by i. When all are left on deposit until the time of the last payment, the level annual payments of X accumulate to \$37,804.39; so $Xs_{\overline{15}|i} = \$37{,}804.39$. On the other hand, the value of the annuity at the time of the inheritance was $Xa_{\overline{15}|i} = \$15{,}077.10$. Therefore, by Equation (3.2.8),

$$(1+i)^{15} = \frac{s_{\overline{15}|i}}{a_{\overline{15}|i}} = \frac{\$37{,}804.39}{\$15{,}077.10} \approx 2.507404607$$

and

$$i = \left[(1+i)^{15}\right]^{1/15} - 1 \approx (2.507404607)^{1/15} - 1 \approx .063199984 \approx 6.32000\%.$$

Moreover,

$$X = \frac{\$37{,}804.39}{s_{\overline{15}|i}} = \frac{\$37{,}804.39i}{(1+i)^{15}-1} \approx \frac{(\$37{,}804.39)(.063199984)}{2.507404607-1} \approx \$1{,}585.000353 \approx \$1{,}585.00.$$

(7) We begin by noting that when the price is low, the buyer needs to borrow less money and hence should be less concerned about the interest rate. In the extreme case where the negotiated price Y is equal to the \$3,600 rebate amount (*not a realistic price*), there would be no money to finance under the rebate option! On the other hand, a high price makes the buyer more sensitive to the interest rate.

If the buyer chooses 0% financing, then the total amount of money to be paid is Y, and since this is spread over sixty level payments, each payment is for an amount $\frac{Y}{60}$. On the otherhand, with the 7% annual effective credit-union financing of the rebate-reduced price $(Y - \$3{,}600)$, the sixty level monthly payments are each for $\frac{Y-\$3{,}600}{a_{\overline{60}|(1.07)^{\frac{1}{12}}-1}}$. In order for the buyer to be indifferent between these two forms of financing, we would have to have these amounts equal; that is, the buyer does not care whether he uses the 0% dealer financing or takes advantage of the rebate and 7% annual effective credit-union financing if

$$\frac{Y}{60} = \frac{Y - \$3{,}600}{a_{\overline{60}|(1.07)^{\frac{1}{12}}-1}}.$$

Equivalently, the methods result in the same payments if

$$Ya_{\overline{60}|(1.07)^{\frac{1}{12}}-1} = 60Y - \$216{,}000,$$

$$216{,}000 = Y\left(60 - a_{\overline{60}|(1.07)^{\frac{1}{12}}-1}\right),$$

or

$$Y = \frac{216{,}000}{60 - a_{\overline{60}|(1.07)^{\frac{1}{12}}-1}}.$$

We calculate that this last expression $\left[216{,}000/60 - a_{\overline{60}|(1.07)^{\frac{1}{12}}-1}\right]$ is approximately 23,830.83044. Now, if the price is above this value and the buyer uses the rebate option, there will be more money borrowed at the high rate of 7%; this is disadvantageous to the buyer so, if the price \$$Y$ is at least \$23,380.84, the buyer would prefer to use the 0% financing.

(3.3) Annuities-due

(1) The annuity has $1984 - 1965 = 19$ annual payments, and we are asked to find the January 1, 1966 value, which is the value at the time of the first payment. Therefore, using an annual effective interest rate of 5%, the January 1, 1966 value is given by the expression $\$3{,}000\ddot{a}_{\overline{19}|5\%}$. By Equations (3.3.5) and (1.9.6), $\$3{,}000\ddot{a}_{\overline{19}|5\%} = \frac{1-(1.05)^{-19}}{1-(1.05)^{-1}} \approx \$38{,}068.76071 \approx \$38{,}068.76$. BA II Plus calculator users, note

that $\$3{,}000\ddot{a}_{\overline{19}|5\%}$ may also be calculated using the **TVM worksheet** in BGN mode with P/Y = C/Y = 1. [To change the mode (if necessary), key 2ND BGN 2ND SET 2ND QUIT. To set P/Y = C/Y = 1 (if necessary), key 2ND P/Y 1 ENTER 2ND QUIT.] Now, push

1 9 N 5 I/Y 3 0 0 0 +/− PMT 0 FV CPT PV,

and your display should show "PMT = 38,068.76071".

(3) (a) The annuity $a_{\overline{n}|i}$ measures the total value one interest period before the first payment of a sequence of n successive end-of-interest-period payments of 1, while the symbol $a_{\overline{n+1}|i}$ gives the value of these n payments of 1, along with an additional payment of 1 at time $n+1$ interest periods. Thus, the difference $a_{\overline{n+1}|i} - a_{\overline{n}|i}$ gives the value of just the additional payment; that is, it gives the value of a payment of 1 to be made in $n+1$ interest periods (where the value is calculated using compound interest with an effective interest rate of i per interest period).

(b) We are given that $\ddot{a}_{\overline{n+1}|i} - \ddot{a}_{\overline{n}|i} = .185248436$, and $a_{\overline{n+1}|i} - a_{\overline{n}|i} = .177208656$. Therefore,

$$.185248436 = \ddot{a}_{\overline{n+1}|i} - \ddot{a}_{\overline{n}|i} = (1+i)a_{\overline{n+1}|i} - (1+i)a_{\overline{n}|i} = (1+i)(a_{\overline{n+1}|i} - a_{\overline{n}|i}) = (1+i)(.177208656);$$

here we have used Equation (3.3.7). So, $1+i = \frac{.185248436}{.177208656}$. Moreover, from the result of part (a), we have $.177208656 = (1+i)^{-(n+1)}$. Therefore,

$$\ln(.177208656) = -(n+1)\ln\left(\frac{.185248436}{.177208656}\right),$$

and

$$n = \left[-\ln(.177208656) \Big/ \ln\left(\frac{.185248436}{.177208656}\right)\right] - 1 \approx 37.99999772.$$

Since the number of payments n is an integer, $n = 38$.

(5) Fred makes $60 - 24 = 36$ annual \$7,500 deposits to the retirement fund. Therefore, at the time of the last deposit, the account has an accumulated value equal to $\$7{,}500 s_{\overline{24}|5\%} \approx \$718{,}772.4204$. Fred, therefore, transfers \$718,772.42 to the account earning an annual effective interest rate j, and his balance five years later is

$$\$718{,}772.42(1+j)^5.$$

We are told that this is just sufficient to cover the purchase of a twenty-five-year annuity-due paying \$5,800 each month, where the purchase price is determined using an annual effective rate equivalent to the monthly effective interest rate $q = (1.04)^{\frac{1}{12}} - 1 \approx .327373978\%$. This annuity has $25 \times 12 = 300$ payments. So,

$$\$718{,}772.42(1+j)^5 \approx \$5{,}800\ddot{a}_{\overline{300}|I} \approx \$1{,}110{,}713.70.$$

It follows that $j \approx \left(\frac{1{,}110{,}713.70}{718{,}772.42}\right)^{\frac{1}{5}} - 1 \approx 9.0943213\% \approx 9.09432\%$.

(7) First recall [Equation (3.3.10)] that $\ddot{s}_{\overline{n}|} + 1 = s_{\overline{n+1}|}$. Therefore,

$$\ddot{s}_{\overline{n}|} = s_{\overline{n+1}|} - 1 = 64{,}024.90944 - 1 = 64{,}023.90944.$$

On the other hand

$$\ddot{s}_{\overline{n}|} = (1+i)^n \ddot{a}_{\overline{n}|} = (1+i)^n(31.61667882).$$

Thus,

$$64{,}023.90944 = (1+i)^n(31.61667882) \quad \text{and} \quad (1+i)^n = \frac{64{,}023.90944}{31.61667882}.$$

It follows that

$$64{,}023.90944 = \ddot{s}_{\overline{n}|} = \frac{(1+i)^n - 1}{d} = \left(\frac{64{,}023.90944}{31.61667882} - 1\right)/d.$$

We therefore find

$$d = \left(\frac{64{,}023.90944}{31.61667882} - 1\right)\Big/ 64{,}023.90944 \approx .031613256$$

and

$$i = \frac{d}{1-d} \approx .03264528 \approx 3.26453\%.$$

To find n, we start with the relationship

$$(1.03264528)^n \approx (1+i)^n = \frac{64{,}023.90944}{31.61667882}$$

and take natural logarithms; we thus obtain

$$n \ln(1.03264528) = \ln(64{,}023.90944/31.61667882), \quad \text{and} \quad n = 237.$$

(9) Consider a loan for an amount 1. If it is to be repaid by n level beginning-of-interest-period payments, the first of which is due immediately, then the amount Q of each level payment satisfies $Q\ddot{a}_{\overline{n}|i} = 1$: thus $Q = \frac{1}{\ddot{a}_{\overline{n}|i}}$.

On the other hand, at the beginning of each interest period for n periods, you might pay the lender the interest d due on the loan of 1 for that period, and then you make no further payments to the lender until time n, when you repay the loan amount 1 (which has not grown since you paid the interest for each period at the beginning of that period.) To accumulate the loan amount 1 at time n, you might make level deposits at the beginning of each of the n periods to an account with a rate identical to that of the loan (effective rate i); these should each be for an amount $R = \frac{1}{\ddot{s}_{\overline{n}|i}}$. Under this second repayment method, your total payment at the beginning of each interest period would be $d + \frac{1}{\ddot{s}_{\overline{n}|i}}$.

Since there is a single interest rate used for either repayment method, the level amounts paid by each method must be the same. So, $d + \frac{1}{\ddot{s}_{\overline{n}|i}} = \frac{1}{\ddot{a}_{\overline{n}|i}}$.

(3.4) Perpetuities

(1) Starting at age forty-five, Kalen (or his heirs) will receive a perpetuity-due with annual payments of \$40,000. Since this sequence of payments may be thought of as a first payment of \$40,000, followed by an annuity-immediate with annual payments of \$40,000, the value of the perpetuity-due on Kalen's forty-fifth birthday, figured using the specified 4% annual effective rate, is $\$40{,}000 + \frac{\$40{,}000}{.04} = \$1{,}040{,}000$. This same answer may be found by computing $\frac{\$40{,}000}{d} = \frac{\$40{,}000}{.04/1.04} = 1.04\left(\frac{\$40{,}000}{.04}\right)$.

(3) (a) The balance n years after the \$40,000 was invested was $\$40{,}000(1.07)^n$, and this must be *at least* $\frac{\$5{,}000}{.07}$ in order for it to fund annual \$5,000 payments forever, beginning $n+1$ years after January 1, 1980. Therefore, we seek the the *smallest* integer n so that

$$\$40{,}000(1.07)^n \geq \frac{\$5,000}{.07},$$

or equivalently, the *smallest* integer n so that $(1.07)^n \geq \frac{5{,}000}{.07(40{,}000)} = \frac{1}{.56}$. The natural logarithm is an increasing function, so this inequality is equivalent to the inequality $n\ln(1.07) \geq -\ln(.56)$. So, we seek the *smallest integer* n with

$$n \geq \frac{-\ln .(56)}{\ln(1.07)} \approx 8.569761713.$$

Thus $n = 9$, and the first \$5,000 scholarship is awarded $n + 1 = 9 + 1 = 10$ years after January 1, 1980; it is awarded on January 1, 1990.

(b) On January 1, 1989, the invested \$40,000 has grown to $\$40{,}000(1.07)^9 \approx \$73{,}538.3685$ and $\frac{\$5{,}000}{.07} \approx$ \$71,428.5714 of this money is needed to support the annual \$5,000 scholarships that begin on January 1, 1990. The balance is $\$73{,}538.3685 - \$71{,}428.5714 \approx \$2{,}109.797$. Therefore, there will be sufficeient money for the \$5,000 scholarships into perpetuity if a \$2,109.79 scholarship is awarded on January 1, 1989, but any larger scholarship would eventually lead to a short-fall.

(5) By Equation (3.4.3), we have $1{,}000 = 54\ddot{a}_{\overline{\infty}|i} = \frac{54}{d}$. Therefore, $d = .054$, $i = \frac{.054}{1-.054}$, and

$$s_{\overline{22}|i} = \frac{(1+i)^{22} - 1}{i} = \frac{(1-d)^{-22} - 1}{i} = \frac{(1-.054)^{-22} - 1}{.054/(1-.054)} \approx 41.89597.$$

(3.5) Deferred annuities and annuity values on any date

(1) The present value of the deferred perpetuity, figured using a compound interest accumulation function $a(t) = (1+i)^t$, is $\frac{\$10{,}000}{i}(1+i)^{-3}$; this follows from the fact that the perpetuity may be viewed as a perpetuity-immediate with annual \$10,000 payments, deferred for three years. When $i = 5\%$, we calculate this to be approximately \$172,767.52, and it is about \$139,936.55 when $i = 6\%$. So, based on compound interest at an annual effective interest rate of 5%, the deferred perpetuity has the larger present value, while the single immediate payment of \$160,000 has the larger present value if you use an annual effective interest rate of 6%.

Sydney should not necessarily make her decision based exclusively on a comparison of the present values. She may have an immediate use (or desire) for the money that makes it particularly desirable to receive the money immediately, and she may deem it better not to borrow, either due a rational analysis of the available loans or due to a personal aversion to borrowing. If she receives the money immediately, she is apt to have a large income tax liability; this should be weighed against the taxes she and her heirs might pay if her award is paid into perpetuity, although these are naturally uncertain. There are other uncertainties Sydney should consider; these include the fact that it is unknown how investments will grow as time goes by and the risk of default by the provider of the perpetuity.

(3) Denote the unknown level payment amount by X. Then the value of the annuity one year before the first payment, calculated using compound interest at an annual effective interest rate of 7.8%, is $Xa_{\overline{12}|7.8\%}$. Thus, figured again using the 7.8% annual effective interest rate, the present value is $(1.078)^{-11}Xa_{\overline{12}|7.8\%}$. We are given that, to the nearest penny, this is equal to \$21,092.04. Thus, $X \approx \frac{\$21{,}092.04}{(1.078)^{-11}a_{\overline{12}|7.8\%}} \approx \$6{,}328.000374$. So, the level payment amount is \$6,328.00.

(5) (a) Let j denote the effective interest rate for half-a-month. There are eleven payments of \$4,166.67, followed by one for four cents less. At the time of the last slightly-reduced payment, the sequence of twelve payments has an accumulated value equal to $\$4{,}166.67s_{\overline{12}|j} - \$.04$. With $i = .06$, $j = (1.06)^{\frac{1}{24}} \approx .243082084\%$, and

$$\$4{,}0166.67s_{\overline{12}|j} - \$.04 \approx \$50{,}673.92249 \approx \$50{,}673.92.$$

(b) There are five-and-a-quarter years between September 30, 1974 and January 1, 1980. Therefore, the January 1, 1980 accumulated value of the twelve deposits was $[\$4{,}0166.67s_{\overline{12}|j} - .04](1.06)^{5.25} \approx$ \$68,808.21877. Since this is to fund an annuity with twenty annual payments, the first of which is on January 1, 1980, we set this equal to $P\ddot{a}_{\overline{20}|6\%}$ and solve for P. We find $P \approx \$5{,}659.447236 \approx$ \$5,659.45. (Rounding up we would have a slight shortfall in the last payment, so you might prefer to round down to \$5,659.44, in which case there is a bit of extra money.)

(3.6) Outstanding loan balances

(1) (a) Immediately after the sixth payment, Jemeel is obligated to make four annual payments of \$1,516 the first of which is due in exactly one year. Since the loan was made at an annual effective interest rate of 6%, by the perspective method, Jemeel's loan balance is $\$1{,}516a_{\overline{4}|6\%} \approx \$5{,}253.10$.

(b) Just after the third payment, seven payments remain; these are three \$822 payments followed by four \$1,516 payments. An easy way to calculate the balance is to use the idea of "imaginary payments"; specifically (from Jemeel's perspective) seven annual \$1,516 outflows remain, along with annual inflows of \$1,516 – \$822 = \$694 for the next three years. So, by the perspective method, the balance just after the third payment is

$$\$1{,}516a_{\overline{7}|6\%} - \$694a_{\overline{3}|6\%} \approx \$8{,}462.890262 - \$1{,}8550.070293 \approx \$6{,}607.81997 \approx \$6{,}607.82.$$

You can also compute this balance by calculating $\$822a_{\overline{3}|6\%} + \$1{,}516a_{\overline{4}|6\%}(1.06)^{-3}$.

(3) Mr. Bell's loan has 15 × 12 = 180 payments, the last 10 × 12 = 120 of which are made when the annual effective interest rate is 6%. The corresponding *monthly* effective interest rate is $j = (1.06)^{\frac{1}{12}} \approx$ .486755057% . Since there is a level payment amount of \$1,692, the balance just after the 100th payment, when 80 payments remain, is $\$1{,}692a_{\overline{80}|j} \approx \$111{,}894.78$. So, since Mr. Bell sold the house for \$258,000, he receives \$258,000 – \$111,894.78. ≈ \$146,105.22.

Note that we used the perspective method here and did not need to consider the initial 4% annual effective interest rate. The retrospective method would have been more cOmplicateded and both annual effective interest rates would have been needed.

(5) (a) The amount financed is \$18,300 – \$3,800 = \$14,500, and the loan has 6 × 12 = 72 payments. The monthly effective interest rate equivalent to the given 5.2% annual effective rate is $j = 1.052)^{\frac{1}{12}} \approx$ .423336166% . Therefore, we compute P such that $Pa_{\overline{72}|j} = \$14{,}500$; we find $P \approx \$234.0602097$. Should we round the payment value of each of the seventy-two payments down to \$234.06, which we are tempted to do since the computed P is only slightly more than this value, we would find that Alice falls one penny short of repaying the loan. So, to find the amount of each of the first 71 payments, we round \$234.0602097 up to \$234.07.

(b) We compute the outstanding loan balance using the retrospective method. Had there not been any payments made, the original \$14,500 balance would have grown to $\$14{,}500(1.052)^2 = \$16{,}047.208$ at the end of two years (twenty-four months). Twenty-four (on-schedule) end-of-month payments reduce the amount owed by $\$234.07s_{\overline{24}|j} \approx \$5{,}899.8515$. So, the balance is about \$16,047.208 – \$5,899.8515 ≈ \$10,147.36.

(7) The monthly effective interest rate is $\frac{4.8\%}{12} = .4\%$ and the loan has 2 × 12 = 24 monthly payments. We therefore look for P with $Pa_{\overline{24}|.4\%} = \$14{,}058.97$, the amount borrowed by Janelle. This yields $P \approx \$635.2299231$. Rounding up to \$635.23, we observe that twenty-four payments of \$635.23 would repay about \$14,508.97176 ≈ \$14,508.97, so there is no need for a slightly reduced final payment. At the end of nine months, without any payment having been made at that time, the outstanding loan balance is $\ddot{a}_{\overline{24-8}|.4\%} \approx \$9{,}865.558365$. If this amount is to be repaid by 24 – 9 = 15 level payments, each should be for an amount Q where $Qa_{\overline{15}|.4\%} = \$9{,}865.558365$. It follow that they should be for $Q = \$678.95$. (This will actually repay \$9,865.61 so, if the new payments are not required to be level, the final one could be reduced by five cents.)

(3.7) Nonlevel annuities

(1) The monthly effective interest rate is $\frac{3\%}{12} = .25\%$, and the accumulated value at the end of twelve months resulting from the end-of-month payments is $\$300 s_{\overline{12}|.25\%}$. At he end of two more years, this would grow to $\$300 s_{\overline{12}|.25\%}(1.0025)^{24} \approx \$3{,}875.322781$. He thus needs an additional $\$14{,}000 - \$3{,}875.322781 = \$10{,}124.67722$. We thus seek the smallest amount Q that is an integer number of cents and satisfies $Q s_{\overline{24}|.25\%} > \$10{,}124.68$. The equation $Q s_{\overline{24}|.25\%} = \$10,124.67722$ yields $Q \approx \$409.8592047$, so we take $Q = \$409.86$.

(3) Let j denote the monthly effective interest rate. Then $(1+j)^{12n} = (1+i)^n \approx 92.372$. We know that $\$217{,}593.30 = \$100 s_{\overline{5n}|j} + \$200 s_{\overline{3n}|j}$. That is,

$$\$217{,}593.30 = \$100\left(\frac{(1+j)^{5n}-1}{j}\right) + \$200\left(\frac{(1+j)^{3n}-1}{j}\right).$$

So, $j \approx \frac{1}{217{,}593.30}\left(100\left[(92.372)^{\frac{5}{12}}\right] + 200\left[(92.372)^{\frac{3}{12}}\right]\right) \approx .004499999$. It follows that $n \approx \frac{\ln(92.372)}{\ln(1+i)} \approx 84.00001283$. Assuming n is an integer, it is 84.

(5) The information given about Sean's repayment will allow us to determine the common amount L that the two boys each borrow from their parents. Note that Sean makes monthly payments, and the effective monthly rate of interest is $j = (1.05)^{\frac{1}{12}} \approx .407412378\%$. Therefore,

$$L = \$500 a_{\overline{96}|j} - \$300 a_{\overline{96}|j} \approx \$39{,}660.13984 - \$13{,}055.38574 \approx \$26{,}604.7541.$$

So, the loan amount is $26,604.7541. [You might find it conceptually easier to compute the amount L Sean repays by using the formula $L = \$200 a_{\overline{48}|j} + \$500 a_{\overline{48}|j}(1.05)^{-4}$, but this probably takes slightly longer than the formula we used, a formula that requires you to view each $200 payment as a $500 payment with a $300 rebate.]

We next find the interest rate for the the period $[0, T]$ (or any period of lenght T) using the **Cash Flow worksheet**; it is also possible to do this using "guess and check" or Newton's method, but that is more time consuming. On your BA II Plus calculator, open and clear the **Cash Flow worksheet** by pressing

[2ND] [CF] [2ND] [CLR WORK].

Then enter the negative of the dollar amount of the loan, $-26,604.7541$, by pushing the sequence

[2] [6] [6] [0] [4] [•] [7] [5] [+/−] [ENTER].

Next key

[↓] [7] [0] [0] [0] [ENTER] [↓] [3] [ENTER] [↓] [3] [0] [0] [0] [ENTER] [↓] [2] [ENTER]
[↓] [0] [ENTER] [↓] [↓] [2] [0] [0] [0] [ENTER]

to enter C01 = 7,000, F01 = 3, C02 = 3,000, F02 = 2, C03 = 0, F03 = 1, C04 = 2,000, and F04 = 1. Follow this with [IRR] [CPT] to display "IRR = 3.104842612". So, the effective interest rate j for the interval $[0, T]$ satisfies $j \approx 3.104842612\%$. Recalling that the annual effective interest rate is 5%, we have

$$(1.05)^T \approx 1.03104368332.$$

It follows that $T \approx \frac{\ln(1.03104368332)}{\ln(1.05)} \approx .626592972 \approx .62659$.

(7) (a) First observe that a nominal interest rate of 8% convertible monthly is equivalent to an annual effective interest rate $i = \left(1 + \frac{.08}{12}\right)^{12} - 1 \approx 8.299950681\%$ and to a two-year effective rate $j = \left(1 + \frac{.08}{12}\right)^{24} - 1 \approx 17.28879318\%$.

Next, think of Lucy's annuity as the sum of two annuities, an annuity paying \$500 on each of her first twenty-one birthdays and an annuity that pays \$200 on her first ten even-numbered birthdays. It follows that the value of the given annuity on the day of Lucy's birth is $\$500a_{\overline{21}|i} + \$200a_{\overline{10}|j}$. This equals approximately $\$4{,}895.113685 + \$922.0176368 \approx \$5{,}817.31$.

(b) Now, we seek the numerical value of the annual effective discount rate d such that $\$6{,}000 = \$500a_{\overline{21}|i} + \$200a_{\overline{10}|j}$; here, $i = (1-d)^{-1} - 1$ is the annual effective interest rate and $j = (1-d)^{-2} - 1$ is the effective interest rate for a two-year period. Let's go ahead and estimate i and then calculate the equivalent d. From part (a), since the value at the time of Lucy's birth is lower than it would be if the annual effective interest rate were about 8.3%, , a "guess and check" solution might reasonably begin with the initial estimate $i = 7\%$. But with $i = 7\%$, Lucy's annuity would have value

$$\$6{,}000 = \$500a_{\overline{21}|i} + \$200a_{\overline{10}|j} \approx \$5{,}417.764 + \$1{,}023.576 \approx \$6{,}441.34.$$

This is too large, so we search for a rate between 7% and 8.3%. Trying $i = 7.9\%$, we obtain

$$\$5{,}047.100 + \$951.577 \approx \$5{,}998.68,$$

so i is just below 7.9%. In fact, as may be seen by further guess and check, or more quickly using the BA II Plus calculator **Cash Flow worksheet**, $i \approx 7.897156752\%$ (This is more accurate than we need to answer the question posed.) The equivalent discount rate $d = 1 - (1+i)^{-1}$ is about 7.319151857%. To the nearest hundredth of a percent, the answer is $d = 7.32\%$.

(3.8) Annuities with payments in geometric progressions

(1) Al's annuity has a first payment of \$100 and then each payment is .96 times the previous one, so the k-th payment is $\$100(.96)^{k-1}$. Since these are annual end-of-year payments, the first of which is paid to Al one year after the date he is given the annuity, if we calculate the value of the k-th payment at the time Al is awarded the annuity, using an annual effective interest rate of 5%, we obtain $\$100(.96)^{k-1}(1.05)^{-k}$. Since there are fifteen payments, figured at the time of the award (*again with* $i = 5\%$) the annuity has value

$$\sum_{k=1}^{15} \$100(.96)^{k-1}1.05^{-k} = \frac{\$100}{1.05} + \left(\frac{\$100}{1.05}\right)\left(\frac{.96}{1.05}\right) + \left(\frac{\$100}{1.05}\right)\left(\frac{.96}{1.05}\right)^2 + \cdots + \left(\frac{\$100}{1.05}\right)\left(\frac{.96}{1.05}\right)^{14}.$$

Using Equation (3.2.2) for the sum of a geometric series, with initial term $c = \frac{\$100}{1.05}$, multiplying factor $r = \frac{.96}{1.05}$, and the number of terms $n = 15$, this series has sum

$$c\left(\frac{1-r^n}{1-r}\right) = \left(\frac{\$100}{1.05}\right)\left[1 - \left(\frac{.96}{1.05}\right)^{15}\right] \Big/ \left(1 - \frac{.96}{1.05}\right) \approx \$821.3857585 \approx \$821.39.$$

We are told that Sal's annuity has the same value as Al's when an annual effective interest rate of 5% is used to compute the values. So, the value of Sal's annuity one year before the first payment is about \$821.3857585. Therefore, the value of Sal's annuity at the time of the last payment is about $\$821.3857585(1.05)^n$. But we know that the value of Sal's annuity at the time of the last payment is \$1,626.29. From

$$\$821.3857585(1.05)^n \approx \$1{,}626.29,$$

we obtain

$$n \approx \ln\left(\frac{1{,}626.29}{821.3857585}\right) / \ln 1.05 \approx 14.0003001.$$

Since n is the number of payments, it is an integer, and thus $n = 14$.

(3) Let $j = (1.04)^{\frac{1}{12}} - 1$, the effective monthly interest rate equaivalent to a 4% annual effective rate of interest.

In the k-th year, for $k \in \{1, 2, \ldots 20\}$, there are payments for the amount $\$1{,}000(1.02)^{k-1}$ at the end of each month. Figured using $i = 4\%$, the accumulated value at the end of the k-th year of the payments occurring that year is $\$1{,}000(1.02)^{k-1} s_{\overline{12}|j}$. To find the present value of the annuity, for each $k \in \{1, 2, \ldots 20\}$, we must bring this value back to time zero and then we must add the twenty time-zero values. Thus, the present value of our annuity is

$$\sum_{k=1}^{20} \$1{,}000(1.02)^{k-1} s_{\overline{12}|j}(1.04)^{-k} = \$1{,}000 s_{\overline{12}|j} \sum_{k=1}^{20} (1.02)^{k-1}(1.04)^{-k}.$$

The series $\sum_{k=1}^{20} (1.02)^{k-1}(1.04)^{-k}$ is geometric with first term $\frac{1}{1.04}$, multiplying factor$\left(\frac{1.02}{1.04}\right)$, and twenty terms, so Equation (3.2.2) tells us that its sum is

$$\frac{1}{1.04}\left[1 - \left(\frac{1.02}{1.04}\right)^{20}\right] \Big/ \left[1 - \left(\frac{1.02}{1.04}\right)\right].$$

This may be simplified as $\left(1 - \left(\frac{1.02}{1.04}\right)^{20}\right) / .02$. Consequently, the present value of our annuity is

$$\$1{,}000 s_{\overline{12}|j}\left[1 - \left(\frac{1.02}{1.04}\right)^{20}\right] / .02 = \$50{,}000 s_{\overline{12}|j}\left[1 - \left(\frac{1.02}{1.04}\right)^{20}\right] \approx \$196,614.8974 \approx \$196{,}614.90.$$

(5) Payments occur on July 1, beginning in 1988, and those occuring in 1988 through 1995 may be valued on January 1, 1996 using just the 6% annual effective interest rate. Since the initial payment is for \$3,000 and then payments increase by a factor of 1.03 each year, the total January 1996 value of these first eight payments is

$$\$3{,}000(1.06)^{7.5} + \$3{,}000(1.03)(1.06)^{6.5} + \$3{,}000(1.03)^{2}(1.06)^{4.5} + \cdots + \$3{,}000(1.03)^{6}(1.06)^{.5}.$$

This is a geometic series, and by Equation (3.2.2), it has sum

$$\$3{,}000(1.06)^{7.5}\left[1 - \left(\frac{1.03}{1.06}\right)^{8}\right] \Big/ \left[1 - \frac{1.03}{1.06}\right] = \$100{,}000(1.06)^{8.5}\left[1 - \left(\frac{1.03}{1.06}\right)^{8}\right] \approx \$31{,}768.62306.$$

The January 1, 1995 value of the payments occurring in 1996 or later may be found by computing the value of these on January 1, 1996 using the 4% annual effective interest rate and then dividing this answer by 1.06; these payments hence add

$$\frac{1}{1.06}\left(\$3{,}000(1.03)^{8}(1.04)^{-.5} + \$3{,}000(1.03)^{9}(1.04)^{-1.5} + \$3{,}000(1.03)^{10}(1.04)^{-2.5} + \cdots\right)$$

to the total value. To calculate this sum, we note that if $|r| < 1$, then can use Equation (3.2.2) to obtain the sum of an infinite geometric series; in fact,

$$\begin{aligned} c + cr + cr^2 + \cdots &= \lim_{n\to\infty} c + cr + cr^2 + \cdots + cr^{n-1} \\ &= c\left(\frac{1 - r^n}{1 - r}\right) \\ &= \frac{c}{(1 - r)}. \end{aligned}$$

So, the payments occuring after 1995 contribute an additional

$$\frac{1}{1.06}\left(\$3{,}000(1.03)^8(1.04)^{-.5}\right)\Big/\left(1-\frac{1.03}{1.04}\right)=\frac{1}{1.06}\left(\$300{,}000(1.03)^8(1.04)^{.5}\right)\approx \$365{,}619.9263$$

to the value. So, the total value is about $\$31{,}768.62306+\$365{,}619.9263=\$397{,}388.5493\approx\$397{,}388.55$.

(3.9) Annuities with payments in arithmetic progressions

(1) (a) The symbol $(Ds)_{\overline{28}|}$ gives the value at the time of the last payment of an annuity with payments at the end of 28 successive interest periods, the k-th of which is for an amount $(29-k)$; that is, the first one is for an amount 28, and subsequent payments are each 1 less than their predecessor.

Figured using an effective interest rate of 3% per interest period, its value is

$$(1.03)^{28}(Da)_{\overline{28}|3\%}=(1.03)^{28}\left(\frac{28-a_{\overline{28}|3\%}}{.03}\right)\approx(2.287927676)\left(\frac{28-18.76410823}{.03}\right)$$
$$\approx 704.3684132\approx 704.37;$$

here, we have used Equation (3.9.7).

(b) The symbol $(I\ddot{a})_{\overline{\infty}|}$ gives the value at the time of the first payment of a perpetuity having its k-th payment be for an amount k; so, the initial payment is for an amount 1 and each subsequent payment is for an amount that is 1 more than its predecessor. The payments occur at intervals of one interest period; you could alternatively express this by saying they occur at the beginning of each interest period.

Figured using an effective interest rate of 3% per interest period, its value is $(1.03)(Ia)_{\overline{\infty}|}=(1.03)\left(\frac{1}{.03}+\frac{1}{(.03^2)}\right)\approx 1.178.78$; here, we have used Equation (3.9.16) with $P=Q=1$.

(c) The symbol $(I_{100,10}a)_{\overline{15}|}$ represents the value one interest period before the first payment of an annuity with payments in fifteen successive interest payments, the amount of the k-th payment being $90+10k$; so, the first payment is for 100 and each subsequent payment is for 10 more than its predecessor. The payments occur at the end of each interest period.

Figured using an effective interest rate of 3% per interest period, using Equation (3.9.4), we see that its value is

$$100a_{\overline{15}|3\%}+\frac{10}{.03}\left(a_{\overline{15}|3\%}-15(1.03)^{-15}\right)\approx \$1{,}193.793509+333.3333333(11.993793509-9.627929211)$$
$$\approx 1{,}963.795467\approx 1{,}963.80.$$

(3) The amount of the k-th payment is $\$290+\$30k$. We note that $\$290+\$30k=\$980$ when $k=\frac{980-230}{30}=23$. So, we seek $(I_{320,30}\ddot{a})_{\overline{23}|4\%}+(1.04)^{-22}\,\$980a_{\overline{\infty}|4\%}$, the sum of the present value of an annuity-due with 23 payments and the value 23 periods before the first payment of a level perpetuity with payments of \$980. But

$$(I_{320,30}\ddot{a})_{\overline{23}|4\%}+(1.04)^{-22}\,\$980a_{\overline{\infty}|\%}=(1.04)(I_{320,30}a)_{\overline{23}|4\%}+(1.04)^{-22}\$980\left(\frac{1}{.04}\right)$$
$$=(1.04)\left[\$320a_{\overline{23}|4\%}+\frac{30}{.04}\left(a_{\overline{23}|4\%}-23(1.04)^{-23}\right)\right]+\$24{,}500(1.04)^{-22}$$
$$=(\$332.80+\$780)a_{\overline{23}|4\%}+(-\$17{,}250+\$24{,}500)(1.04)^{-22}$$
$$\approx \$16{,}532.69341+\$3{,}059.176553$$
$$\approx \$19{,}591.86996\approx \$19{,}591.87.$$

(5) Call the six times Bob makes an \$11,000 deposit times 0, 1, 2, 3, 4, and 5. Then, for $k \in \{1, 2, 3, 4, 5, 6\}$, the balance in Bob's 7.5% fund at the beginning of the k-th year (just after any interest has been paid out) is \$11,000k. Moreover, because Bob has no contributions to the 7.5% after the sixth year, the balance in Bob's 7.5% account at the beginning of years 7 through 13 (just after interest has been paid out) is $6 \times$ \$11,000 = \$66,000. Note that the annual interest on \$11,000, calculated using an annual effective rate of 7.5%, is $.075 \times$ \$11,000 = \$825. Therefore, the 5% account receives six increasing deposits of \$825, $2 \times$ \$825, $3 \times$ \$825, $4 \times$ \$825, $5 \times$ \$825, and $6 \times$ \$825 = \$4,950, at times 1,2,3,4,5, and 6, followed by seven level deposits of \$4,950 at times 7 through 13. So, the balance in the 5% account at time 13 is

$$\$825(Is)_{\overline{6}|5\%}(1.05)^7 + \$4{,}950 s_{\overline{7}|5\%} \approx \$18{,}843.13948(1.05)^7 + \$40{,}302.94184 \approx \$66{,}817.13.$$

The total liquidated amount is \$66,000 + \$66,817.13 ≈ \$132,817.13.

(7) Approach 1 (*uses the hint*): Denote the present value of the annuity which lasts n interest periods and pays k^2 at the end of the k-th period by A. Then

$$A = 1^2v + 2^2v^2 + 3^2v^3 + 4^2v^4 + \cdots + n^2v^n,$$

and

$$Av = 1^2v^2 + 2^2v^3 + 3^2v^4 + + \cdots + (n-1)^2v^n + n^2v^{n+1}.$$

So,

$$\begin{aligned}(1-v)A &= A - Av \\ &= \left(1^2v + 2^2v^2 + 3^2v^3 + 4^2v^4 + \cdots + n^2v^n\right) - \left(1^2v^2 + 2^2v^3 + 3^2v^4 + + \cdots + (n-1)^2v^n + n^2v^{n+1}\right) \\ &= 1^2v + (2^2 - 1^2)v^2 + (3^2 - 2^2)v^3 + (4^2 - 3^2)v^4 + \cdots + \left(n^2 - (n-1)^2v^n\right) - n^2v^{n+1}.\end{aligned}$$

Repeatedly using the formula $a^2 - b^2 = (a-b)(a+b)$ for the difference of two squares, we may rewrite this as

$$(1-v)A = 1v + 3v^2 + 5v^3 + 7v^4 + \cdots + (2n-1)v^n - n^2v^{n+1}.$$

Next recall that according to Equations (1.9.4) and (1.9.7), $1 - v = d = \frac{i}{1+i}$. Consequently,

$$\begin{aligned}A &= \left(\frac{1+i}{i}\right)\left[1v + 3v^2 + 5v^3 + 7v^4 + \cdots + (2n-1)v^n - n^2v^{n+1}\right] \\ &= \frac{1}{i}\left[1 + 3v + 5v^2 + 7v^3 + (2n-1)v^{n-1} - n^2v^n\right].\end{aligned}$$

The problem asks you to write A in terms of the annuity symbol $(I_{3,2}a)_{\overline{n-2}|i}$. We note that

$$(I_{3,2}a)_{\overline{n-2}|i} = 3v + 5v^2 + 7v^3 + \cdots + [2(n-2) + 1]v^{n-2}.$$

Therefore,

$$A = \frac{1}{i}[1 + (I_{3,2}a)_{\overline{n-2}|i} + (2n-1)v^{n-1} - n^2v^n].$$

An even simpler formula would be

$$A = \frac{1}{i}[1 + (I_{3,2}a)_{\overline{n-1}|i} - n^2v^n].$$

Approach 2 : We are looking at an annuity whose k-th payment is k^2. Observe that

$$k^2 = \frac{1}{2}[1+3+5+\cdots+(2k-1)].$$

This can be checked by induction or more simply by noting that if $S = 1+3+5+\cdots+(2k-1)$, the sum of the first k integers, then

$$\begin{aligned}2S &= [1+3+5+\cdots+(2k-3)+(2k-1)] + [(2k-1)+(2k-3)+(2k-5)+\cdots+3+1]\\ &= [1+(2k-1)]+[3+(2k-3)]+[5+(2k-5)]+\cdots+[(2k-3)+3]+[(2k-1)+1] = k(2k) = 2k^2.\end{aligned}$$

Therefore, the present value A of the given annuity is

$$\begin{aligned}A &= 1^2v + 2^2v^2 + 3^2v^3 + \cdots + n^2v^n\\ &= 1v + (1+3)v^2 + (1+3+5)v^3 + \cdots + [1+3+5+\cdots+(2n-1)]v^n\\ &= [1v+1v^2+1v^3+\cdots+1v^n] + [3v^2+3v^3+\cdots+3v^n] + [5v^3+5v^4+\cdots+5v^n] + \cdots + (2n-1)v^n\\ &= a_{\overline{n}|i} + 3va_{\overline{n-1}|i} + 5v^2a_{\overline{n-2}|i} + (2n-1)v^na_{\overline{n}|i}\\ &= \left[\frac{1-v^n}{i}\right] + 3v\left[\frac{1-v^{n-1}}{i}\right] + 5v^2\left[\frac{1-v^{n-2}}{i}\right] + \cdots (2n-1)v^n\left[\frac{1-v}{i}\right].\\ &= \frac{1}{i}\left(\left[1+3v+5v^2+\cdots+(2n-1)v^n\right] - v^n\left[1+3+\cdots(2n-1)\right]\right)\\ &= \frac{1}{i}\left(1+(I_{3,2}a)_{\overline{n-1}|}) - v^nn^2\right)\\ &= \frac{1}{i}\left[1+(I_{3,2}a)_{\overline{n-2}|i} + (2n-1)v^{n-1} - n^2v^n\right].\end{aligned}$$

Numerical solution: if $n = 30$ and $i = .04$, then

$$\begin{aligned}(I_{3,2}a)_{\overline{n-1}|} &= 3a_{\overline{n-1}|i} + \frac{2}{i}\left[a_{\overline{n-1}|i} - (n-1)v^{n-1}\right]\\ &= 3a_{\overline{29}|4\%} + \frac{2}{i}\left[a_{\overline{29}|4\%} - 29)(1.04)^{-29}\right]\\ &\approx 50.9511439 + 50(16.98371463 - 9.298891026) \approx 435.1923242,\end{aligned}$$

and

$$A = \frac{1}{i}\left((1+(I_{3,2}a)_{\overline{n-1}|}) - v^nn^2\right] \approx \frac{1}{.04}\left[1+435.1923242 - (30)^2(1.04)^{-30}\right] \approx 3{,}967.63808.$$

(3.10) Yield rate examples involving annuities

(1) As with any yield rate problem, it is good to start by making a list of cashflows exchanged. Denoting the time of the \$58,000 investment as time 0 and taking the perspective of the investor, the cashflows consist of contributions of −\$58,000 at time 0, \$7,000 − \$4,000 = \$3,000 at times $1, 2, \ldots, 12$, and \$15,000 + \$4,000$\ddot{s}_{\overline{12}|9\%}$ = \$11,000 + \$4,000$s_{\overline{13}|9\%}$ ≈ \$102,813.5383 at time 13. Consequently, a time 0 equation of value for this financial situation is

$$\$58{,}000 = \$3{,}000a_{\overline{12}|i} + (\$11{,}000 + \$4{,}000s_{\overline{13}|9\%})(1+i)^{-13}.$$

You could then use "guess and check" or Newton's method to find $i \approx 8.4326019097\%$, but it is easier to use the **Cash Flow worksheet** with CF0 = −58,000, C01 = 3,000, F01=12, C02 = 102,813.5383, and F02 = 1 to obtain "IRR = 8.432601909". So, the annual yield rate is about 8.43260%.

(3) The cost of the annuity with \$1,000 annual payments, priced using an annual effective interest rate of 8%, is \$7,246.89 since $\$1{,}000\ddot{a}_{\overline{10}|8\%} \approx \$7{,}246.887911$. Therefore, the time 0 value of the certificate of deposit is $10{,}000 - \$7{,}246.89 = \$2{,}753.11$, and the time 10 maturity value of the certificate of deposit is $\$2{,}753.11(1+\frac{.09}{4})^{40} \approx \$6,704.34$. The \$1,000 annuity payments are each immediately deposited into an account earning an annual effective rate of 7%; this accumulates to $\$1{,}000\ddot{s}_{\overline{10}|7\%} \approx \$14{,}783.59932$ at the end of the ten years. Therefore, the investor's \$10,000 grows to a total of $\$6{,}704.34 + \$14{,}783.59932 = \$21{,}487.93932$ in ten years, and the annual yield rate i satisfies $\$21{,}487.93932 = 10{,}000(1+i)^{10}$. It follows that $i \approx 7.949212109\% \approx 7.94921\%$.

(5) Company A pays a total of \$800,000 now, so as to receive \$700,000 in three years and \$250,000 in four years. We do not need to consider the source of these payments to determine that A has time 0 equation of value

$$\$800{,}000 = \$700{,}000(1+i_A)^{-3} + \$250{,}000(1+i_A)^{-4}.$$

By "guess and check", "Newton's method", or most quickly using the **Cash Flow worksheet**, company A's yield rate i_A satisfies $i_A \approx 5.416159831\% \approx 5.416\%$. These same three methods can be used to find Company B's yield rate i_B and Company C's yield rate i_C from the equations of value

$$\$300{,}000 = \$100{,}000(1+i_B)^{-1} + \$250{,}000(1+i_B)^{-4},$$

and

$$\$600{,}000 = \$700{,}000(1+i_C)^{-4}.$$

In fact, $i_B \approx 5.104761093\% \approx 5.105\%$ and $i_C \approx 5.583786208\% \approx 5.584\%$. (*Note that the book asks for the solutions to the nearest thousandth, hence our final estimates; the answers are listed with more places of accuracy in the back of the book.*)

(7) Denote the time at which Tom invests \$80,000 as time 0. Then, Tom receives net payments (after reinvestment of $\$12{,}000 - \$5{,}000 = \$7{,}000$ at times 1, 2, 3, 4, and 5. At time 6, he receives the accumulated value $\$5{,}000\ddot{s}_{\overline{5}|5\%}$ of his 5% reinvestment account. Therefore, a time 0 equation of value for his investment is

$$\$80,000 = \$7{,}000a_{\overline{5}|i} + \$5{,}000\ddot{s}_{\overline{5}|5\%}(1.05)^{-6}.$$

Note that $\$5{,}000\ddot{s}_{\overline{5}|5\%} \approx \$29{,}009.56406$, so the total amount received by Tom is about

$$(5 \times \$7{,}000) + \$29{,}009.56 \approx \$64{,}009.56$$

which is about \$15,990.43 *less* than he invested. This *loss* takes place over a six year period, so there is an annual loss of about \$2,665. Note that $\frac{\$2{,}665}{\$80{,}000} \approx 3.33\%$ but thanks to compounding, Tom's yield rate will be smaller than -3.33%. In fact, as you may check using "guess and check", Newton', method, the **TVM worksheet,** or the **Cash Flow worsksheet** with CF0 = –80,000, C01 = 7,000, F01 =5, C02 = 29,009.56, and F02 =2, Tom's yield rate is approximately $i = -4.89686\%$.

(9) Let j denote the effective quarterly rate of interest. Then, since there are three months in a quarter, a time 0 equation of value for this financial relationship is

$$\$10{,}000 = \$1{,}000a_{\overline{12}|j} + \$300(1+j)^{-\frac{61}{3}}.$$

Therefore, the interest rate j is a root of the function $f(j) = 1{,}000a_{\overline{12}|j} + 300(1+j)^{-\frac{61}{3}} - 10{,}000$. We find a narrow interval in which j must lie using the "guess and check" method ; j may also be estimated using Newton's method.

You might find an initial (slightly too high) guess for j using the **Cash Flow worksheet** and moving the \$300 payment forward one month to time 60 months (20 quarters); setting CF0 = –10,000, C01 = 1,000, F01 =12, C02 = 0, F02 = 7, C03 = 300, and F03 = 1 ; $j \approx 3.193473297$.

We calculate that $f(.0319) \approx .47006419$, $f(.031905) \approx .16228766$, and $f(.03191) \approx -.145474427$, so $.031905 < j < .03191$. Noting that $(1.031905)^4 - 1 \approx .133858518 \approx 13.39\%$ and $(1.03191)^4 - 1 \approx .133880495 \approx 13.39\%$, $i = (1+j)^4 \approx 13.39\%$.

(3.11) Annuity-symbols for nonintegral terms

(1) The quarrterly effective interest rate is $\frac{4\%}{4} = 1\%$, so we first seek N with

$$\$200{,}0000 = \$25{,}000 a_{\overline{N}|1\%} = \$25{,}000\left(\frac{1-(1.01)^{N}}{.01}\right);$$

the number of payments will be the smallest integer which is at least N. We find $N \approx 8.379.$, so there are nine payments, the first eight of which are for \$25,000. To find the amount X of the ninth payment (the drop payment), solve the equation $\$200{,}0000 = \$25{,}000 a_{\overline{8}|1\%} + X(1.01)^{-9}$;

$$X \approx (1.01)^{9}\$200{,}0000 - \$191{,}291.9438 \approx \$9{,}523.872826.$$

The amount of the drop payment must be an integer number of cents, hence is \$9,523.87.

(3) Since $\$600 a_{\overline{T}|5\%} = \$600\left(\frac{1-(1.05)^{-T}}{.05}\right) = \$12{,}000(1-(1.05)^{-T})$, the equation $\$5{,}000 = \$600 a_{\overline{T}|5\%}$ is equivalent to $(1.05)^{-T} = 1 - \frac{5{,}000}{12{,}000} = \frac{7}{12}$, and the term of the loan is $T = \ln(60/35)/\ln(1.05) \approx 11.0472387 \approx 11.04724$. Next, to find the amount X of the payment at time T, consider the equation $\$5{,}000 = \$600 a_{\overline{11}|5\%} + X(1.05)^{-T}$. Using the just-calculated value of T ($T \approx 11.0472387$), we find

$$X = (\$5{,}000 - \$600 a_{\overline{11}|5\%})(1.05)^{T} \approx \$27.68823263 \approx \$27.69.$$

(3.12) Annuities governed by general accumulation functions

(1) (a) The accumulated value is

$$\begin{aligned}&\$100\frac{a(6)}{a(1)} + \$100\frac{a(6)}{a(2)} + \$100\frac{a(6)}{a(3)} + \$100\frac{a(6)}{a(4)} + \$100\frac{a(6)}{a(5)} + \$100\frac{a(6)}{a(6)}\\ &= \$100a(6)\left(\frac{1}{a(1)} + \frac{1}{a(2)} + \frac{1}{a(3)} + \frac{1}{a(4)} + \frac{1}{a(5)}\ \frac{1}{a(6)}\right)\\ &= \$100(1.24)\left(\frac{1}{1.04} + \frac{1}{1.08} + \frac{1}{1.12} + \frac{1}{1.16} + \frac{1}{1.20}\ \frac{1}{1.24}\right)\\ &\approx \$654.9897548 \approx \$654.99.\end{aligned}$$

(b) We seek i so that $\$654.99 = \$100 s_{\overline{6}|i} = \frac{(1+i)^6-1}{i}$. This may be found by "guess and check" or Newton's method, but it is most easily found using the **TVM worksheet** in END mode with P/Y=C/Y=1; key

[6] [N] [0] [PV] [6] [0] [0] [+/−] [FV] [6] [5] [4] [•] [9] [9] [PMT] [CPT] [I/Y]

to obtain "I/Y = 3.498467526", so $i \approx 3.49847\%$. This differs in the last decimal place from the answer reported in the back of the text; that answer is obtained if you do not round to the nearest cent in part (a) but instead enter the stored value 654.9897548 in the FV register.

(3) The discount function $v(t)$ is $\frac{1}{a(t)} = \frac{3}{(t+1)(t+3)} = \frac{1.5}{t+1} - \frac{1.5}{t+3}$; here, we obtained the latter expression by noting that

$$\frac{3}{(t+1)(t+3)} = \frac{A}{t+1} - \frac{B}{t+3} = \frac{A(t+3)+B(t+1)}{(t+1)(t+3)} = \frac{(A+B)t+(3A+B)}{(t+1)(t+3)}$$

when $A+B=0$ and $3A+B=3$ or equivalently, when $A=1.5$ and $B=-1.5$.
The time-zero value of the perpetuity is

$$\begin{aligned}\sum_{k=3}^{\infty}\$1{,}000v(t) &= \sum_{k=3}^{\infty}\$1{,}000\left(\frac{1.5}{k+1}-\frac{1.5}{k+3}\right)\\ &= \$1{,}000\left[\left(\frac{1.5}{4}-\frac{1.5}{6}\right)+\left(\frac{1.5}{5}-\frac{1.5}{7}\right)+\left(\frac{1.5}{6}-\frac{1.5}{8}\right)+\left(\frac{1.5}{7}-\frac{1.5}{9}\right)+\cdots\right]\\ &= \$1{,}000\left[\frac{1.5}{4}+\frac{1.5}{5}+\left(\frac{1.5}{6}-\frac{1.5}{6}\right)+\left(\frac{1.5}{7}-\frac{1.5}{7}\right)+\left(\frac{1.5}{8}-\frac{1.5}{8}\right)+\cdots\right]\\ &= \$1{,}000\left[\frac{1.5}{4}+\frac{1.5}{5}\right] = 675.\end{aligned}$$

(3.13) The investment year Method

(1) The \$1,000 deposited at the beginning of 1989 contributes $\$1{,}000(1.06)(1.055)(1.05)(1.045)(1.05) \approx \$1{,}2888.407409$ to the January 1, 1994 balance; the first four applicable interest rates are found by looking at the row beginning with "1989", and the rate for 1993 is found in the fourth column of the row which begins with "1990". The \$1,000 deposited at the beginning of 1990 grows to $\$1{,}000(1.065)(1.06)(1.055)(1.05) \approx \$1{,}250.538975$ as of January 1, 1994, and the \$1,000 deposited at the beginning of 1990 accumulates to $\$1{,}000(1.06)(1.055)(1.05) \approx \$1{,}174.215$. So, the total accumulation is $\$1{,}2888.407409 + \$1{,}250.538975 + \$1{,}174.215 = \$3{,}713.161384 \approx \$3{,}713.16$.

Chapter 3 review problems

(1) It is best to think of the annuity as the sum of three annuities, with time t payments of

(1) $\$200(1.03)^{t+1}$,
(2) $\$100(1.02)^{2(t+1)}$,
(3) $\$80(t+1) - 60 = \$80t + \$20$,

respectively. Annuities (1) and (2) each have payment amounts which form geometric progressions, and hence time 10 values may be found using Equation (3.2.2); Figured using an annual effective interest rate of 4%, annuity (1) has time 10 value

$$\begin{aligned}A^{(1)} = \sum_{k=0}^{29}\$200(1.03)^{k+1}(1.04)^{10-k} &= \$206(1.04)^{10}\left(\frac{1-\left[\frac{1.03}{1.04}\right)^{30}}{1-\frac{1.03}{1.04}}\right]\\ &= \$20{,}600(1.04)^{11}\left[1-\left(\frac{1.03}{1.04}\right)^{30}\right]\\ &\approx \$7{,}979.869652\end{aligned}$$

and, since $(1.02)^2 = 1.0404$, annuity (2) has present value

$$A^{(2)} = \sum_{k=0}^{29} \$100(1.0404)^{k+1}(1.04)^{10-k} = \$104.40(1.04)^{10}\left[\frac{\left(\frac{1.0404}{1.04}\right)^{30}-1}{\frac{1.0404}{1.04}-1}\right]$$
$$= \left(\frac{\$104.40}{.0004}\right)(1.04)^{11}\left[\left(\frac{1.0404}{1.04}\right)^{30}-1\right]$$
$$\approx \$4{,}645.997353$$

Annuity (3) has payments in arithmetic progression and has time 10 value

$$A^{(3)} = (1.04)^{10}(I_{20,80}\ddot{a})_{\overline{30}|4\%} = (1.04)^{11}[(I_{20,80}a)_{\overline{30}|4\%}$$
$$= (1.04)^{11}\left[\$20a_{\overline{30}|4\%} + \frac{\$80}{.04}\left(a_{\overline{30}|4\%} - 30(1.04)^{-30}\right)\right]$$
$$= (1.04)^{11}\left[\$2{,}020a_{\overline{30}|4\%} + \$60{,}000(1.04)^{-30}\right]$$
$$\approx (1.04)^{11}(\$34{,}929.90727 - \$18{,}499.12008) \approx \$25{,}294.44199.$$

Therefore, Pierre's annuity, which is the total of the three annuities we have considered, has time 10 value

$$\$7{,}979.869652 + \$4{,}645.997353 + \$25{,}294.44199 \approx \$37{,}920.309 \approx \$37{,}920.31.$$

(3) The present value of Cheryl's annuity is

$$\$6{,}000\ddot{a}_{\overline{60}|5\%} - \$3{,}000\ddot{a}_{\overline{20}|5\%} \approx \$119{,}254.524 - \$39{,}255.96258 \approx \$79{,}998.56143;$$

here we have thought of each of the first twenty \$3,000 annuity payments as if Cheryl received a larger payment of \$6,000 but immediately returned \$3,000. Therefore, David's perpetuity has present value \$79,998.56143, and thinking of his annuity as the sum of a perpetuity with annual end-of-year payments of P along with another perpetuity paying P at the end of two years, we have

$$\$79{,}998.56143 = \frac{P}{.05} + \frac{P}{(1.05)^2 - 1}.$$

It follows that $P \approx \$2{,}688.476245 \approx \$2{,}688.48$.

(5) Starting at the beginning of 1980 and continuing through 1997, an eighteen-year period, the December 31 value of the payments Julie received during the calendar year is $\$100(1+i)^{.25} + \$200$. Therefore, the January 1 , 1980 value of the annuity is

$$\left[\$100(1+i)^{.25} + \$200\right] a_{\overline{18}|i} = \left[\$100(1+i)^{.25} + \$200\right]\left(\frac{1-(1+i)^{-18}}{i}\right);$$

here, we have used Equation (3.2.4). We wish to find the annual effective interest rate i so that this value is \$3,150. Equivalently, we seek i with $f(i) = 0$ where

$$f(i) = \left[100(1+i)^{.25} + 200\right]\left(1-(1+i)^{-18}\right) - 3{,}150i$$
$$= 100(1+i)^{.25} - 100(1+i)^{-17.75} + 200 - 200(1+i)^{-18} - 3{,}150i.$$

The root i may be found by either "guess and check " or Newton's method; an initial guess might reasonably be obtained by pretending that each year's \$300 was paid at the very end of the year, in which case a present value of \$3,150 would correspond to an interest rate of approximately 6.41202502309% (as may easily be checked using the TVM). Of course, getting some of the money earlier would lead to a present value higher than than \$3,150 unless the interest rate used to compute the present value was higher than the just reported rate 6.41202502309%. In fact, $i \approx .064838$, and we will verify this by Newton's method starting with the somewhat less good initial estimate $i_1 = .05$.

Observe that

$$f'(i) = 25(1+i)^{-.75} + 1{,}775(1+i)^{-18.75} + 3{,}600(1+i)^{-19} - 3{,}150.$$

It follows , using the general recipe given by Newton's method (*see page 83 of the text*) and the initial estimate $i_1 = .03$ that

$$i_2 = i_1 - \frac{f(i_1)}{f'(i_1)} \approx .05 - \left(\frac{-18.56109029}{-990.2079029}\right) \approx .06874464,$$
$$i_3 = i_2 - \frac{f(i_2)}{f'(i_2)} \approx .06874464 - \left(\frac{-6.030732833}{-1{,}598.050132}\right) \approx .064970833,$$
$$i_4 = i_3 - \frac{f(i_3)}{f'(i_3)} \approx .064970833 - \left(\frac{-.197323859}{-1{,}492.232876}\right) \approx .064838599,$$
$$i_5 = i_4 - \frac{f(i_4)}{f'(i_4)} \approx .06483599 - \left(\frac{-.001254099}{-1{,}488.38814}\right) \approx .064838428.$$

Moreover, $f(i_5) \approx f(.064838428) \approx 5 \times 10^{-10}$, $f'(i_5) \approx -1{,}488.38317$, and $i_6 = i_5 - \frac{f(i_5)}{f'(i_5)} \approx .06483842828 - \left(\frac{5\times 10^{-10}}{-1{,}488.38317}\right) \approx .064838428 \approx i_5$. So $i \approx .0648384$.

(7) The annuity has two parts. The first part is made up of twenty-four annual payments, the first of which is for \$2,000 and occurs immediately; then each subsequent payment is (1.03) times its predecessor. The second part is a level annuity with $50 - 24 = 26$ payments, the first of which occurs in twenty-four years.

\$2,000	\$2,000(1.03)	$\$2{,}000(1.03)^2$	$\cdots$	$\$2{,}000(1.03)^{23}$	\$4,000	\$4,000 $\cdots$	\$4,000	
0	1	2	$\cdots$	23	24	25 $\cdots$	49	50
PART 1					PART 2			

We want the sum of the values of these parts one period after the final payment of the second part; this is 27 periods after the last payment of the first part (*time* 50 *on our timeline*). Calculated using a 4.2% annual effective interest rate, the sum we seek is

$$\$2{,}000(1.042)^{50} + \$2{,}000(1.03)(1.042)^{49} + \$2{,}000(1.03)^2(1.042)^{48} + \$2{,}000(1.03)^3(1.042)^{47} + \cdots$$
$$+ \$2{,}000(1.03)^{23}(1.042)^{27} + \$4{,}000\ddot{s}_{\overline{26}|4.2\%}$$
$$= \$2{,}000(1.042)^{50}\left[\frac{1-\left(\frac{1.03}{1.042}\right)^{24}}{1-\left(\frac{1.03}{1.042}\right)}\right] + \$4,000\ddot{s}_{\overline{26}|4.2\%}$$
$$= \frac{\$2{,}000}{.012}(1.042)^{51}\left[1-\left(\frac{1.03}{1.042}\right)^{24}\right] + \$4{,}000\ddot{s}_{\overline{26}|4.2\%}$$
$$\approx \$329{,}739.996 + \$189{,}989.0913 \approx \$519{,}729.0873 \approx \$519{,}729.09.$$

CHAPTER 4

Annuities with different payment and conversion periods

(4.2) Level annuities with payments less frequent than each interest period

(1) (a) The effective monthly interest rate is 4.85%/12 and there are three months in a quarter, so the quarterly effective interest rate j is $\left(1+\frac{.0485\%}{12}\right)^3-1\approx 1.217407123\%\approx 1.21741\%$. The balance in the savings account immediately following the last payment is $\$100s_{\overline{40}|j}\approx \$5{,}114.050906\approx \$5{,}114.05$.

(b) The annuity lasts for 120 months and has a payment at the end of each three-month period. Therefore, recalling Equation (4.2.4), we find that the balance immediately after the last payment is $\$100\left(\frac{s_{\overline{120}|4.85\%/12}}{s_{\overline{3}|4.85\%/12}}\right)\approx \$100\left(\frac{154.0424412}{3{,}012141335}\right)\approx \$5{,}114.050906\approx \$5{,}114.05$. The two methods take similar amounts of time; they are both very quick.

(c) You should be in END Mode. Push **2ND** **P/Y** **4** **ENTER** **↓** **1** **2** **ENTER** **2ND** **QUIT** to set your calculator with the desired settings for P/Y and C/Y. Then key **4** **0** **N** **4** **•** **8** **5** **I/Y** **0** **PV** **1** **0** **0** **+/−** **PMT** **CPT** **FV**. Your display should read "FV = 5,114.050906".

NOTE: At this point, you would be wise to return your calculator to your preferred settings for P/Y and C/Y ; these are likely to be P/Y=C/Y=1.

(3) Chapter 3 Method: The interest rate for a two-year period is $(1.06)^2-1=12.36\%$. Therefore, her balance in her savings account is $\$1{,}000\ddot{s}_{\overline{5}|12.36\%}\approx \$7{,}189.291843\approx \$7{,}189.29$. Renee sells the perpetuity at time 10 for $\$1{,}000+\frac{\$1{,}000}{.1236}\approx \$9{,}090.614887\approx \$9{,}090.61$. Therefore, Renee has $\$7{,}189.29+\$9{,}090.61=\$16{,}279.90$ to spend on the thirty year annuity. This annuity has payments at the end of each three years, and the interest rate for a three-year period is $(1.06)^3-1=19.1016\%$. So, $\$16{,}279.90=Pa_{\overline{10}|19.1016\%}$ where P is the desired payment. We calculate $P\approx \$3{,}765.297886\approx \$3{,}765.30$.

Section (4.2) Method: According to Fact (4.2.8), Renee's balance at the end of ten years is

$$\$1{,}000\left(\frac{s_{\overline{10}|6\%}}{a_{\overline{2}|6\%}}\right)\approx \$1{,}000\left(\frac{13.18079494}{1{,}833392666}\right)\approx \$7{,}189.291843\approx \$7{,}189.29.$$

At time 10, Equation (4.2.11) tells us that Renee's price for the perpetuity is calculated to be $\$1{,}000\left(\frac{1}{.06a_{\overline{2}|6\%}}\right)\approx \$1{,}000\left(\frac{1}{.11000356}\right)\approx \$9{,}090.614887\approx \$9{,}090.61$. So, as in part (a), Renee has $\$7{,}189.29+\$9{,}090.61=\$16{,}279.90$ to spend on the thirty year annuity. If this annuity has payments of P at the end of each three years, its value, by Fact (4.2.5) is $P\frac{a_{\overline{30}|6\%}}{s_{\overline{3}|6\%}}\approx P\frac{13.76483115}{3.1836}\approx P(4.323668536)$, so $P\approx\frac{16{,}279.90}{4.323668536}\approx \$3{,}765.297886=\$3{,}765.30$.

(5) Using Fact (4.2.8), we note that $\$4{,}769.30 = \$1{,}000\frac{a_{\overline{72}|3.6575\%}}{a_{\overline{k}|3.6575\%}}$. Therefore,

$$a_{\overline{k}|3.6575\%} = \frac{a_{\overline{72}|3.6575\%}}{4.7693} \approx \frac{25.28252211}{4.7693} \approx 5.301097039.$$

It follows that $N \approx 6.000001867 \approx 6.000002 \approx 6.00000$; this may be determined using Equation (3.2.4) and logarithms, but it is most easily obtained using the **TVM worksheet**. If we assume N is an integer, then we conclude $N = 6$.

(7) Chapter 3 method: The interest rate for a three year period is $(1+i)^3 - 1$, so a perpetuity with level payments of X at the end of each three years has present value $\frac{X}{(1+i)^3-1}$. On the other hand, the interest rate for a two year period is $(1+i)^2 - 1$, so a perpetuity with payments of \$1,000 at the end of each two years has present value $\frac{\$1{,}000}{(1+i)^2-1}$, and an an annuity with payments of \$1,000 at the beginning of each two years has present value $\frac{\$1{,}000}{(1+i)^2-1} + 1$. Therefore,

$$\frac{\$1{,}000}{(1+i)^2 - 1} + 1 = \frac{X}{(1+i)^3 - 1},$$

and

$$X = [(1+i)^3 - 1]\left(\$1{,}000 + \frac{\$1{,}000}{(1+i)^2 - 1}\right).$$

Section (4.2) method: The perpetuity paying \$1,000 at the beginning of each two years has present value $\frac{\$1{,}000}{ia_{\overline{2}|i}} = \frac{\$1{,}000}{1-v^2}$, while the perpetuity with level payments of X at the end of each three years has present value $\frac{X}{is_{\overline{3}|i}} = \frac{X}{(1+i)^3-1}$; here, we have used Equations (4.2.10) and (4.2.11). These present values are equal, so

$$X = \frac{\$1{,}000[(1+i)^3 - 1]}{1 - v^2}.$$

This second answer looks different from the answer we obtained by changing interest rates (the "Chapter 3 method"). However, you may verify that they are equal, since

$$\$1{,}000 + \frac{\$1{,}000}{(1+i)^2 - 1} = \frac{\$1{,}000[(1+i)^2 - 1]}{(1+i)^2} + \frac{\$1{,}000}{(1+i)^2} = \frac{\$1{,}000(1+i)^2}{(1+i)^2 - 1} = \frac{\$1{,}000}{1 - v^2}.$$

(4.3) Level annuities with payments more frequent than each interest period

(1) The annuity symbol $\ddot{s}^{(4)}_{\overline{23}|2.25\%}$ gives the accumulated value at the end of 23 interest periods of an annuity that pays $\frac{1}{4}$ at the beginning of each quarter of an interest period; the effective interest rate for the interest period is 2.25%.

Given an effective interest rate $i = 2.25\%$, the equivalent nominal rate discount rate convertible quarterly is $d^{(4)} = 4\left[1 - (1+i)^{-\frac{1}{4}}\right] \approx .022188837$, and

$$\ddot{s}^{(4)}_{\overline{23}|2.25\%} = \frac{(1.0225)^{23} - 1}{d^{(4)}} \approx \frac{.66823137}{.022188837} \approx 30.11565502 \approx 30.11566.$$

The annuity symbol $a^{(12)}_{\overline{\infty}|4\%}$ gives the present value of a perpetuity with a payment of $\frac{1}{12}$ at the end of each twelfth of an interest period; the effective interest rate for the interest period is 4%.

Given an effective interest rate $i = 4\%$, the equivalent nominal rate interest rate convertible twelve times per year is $i^{(4)} = 12\left[(1+i)^{\frac{1}{12}} - 1\right] \approx 3.928487738$, and

$$a^{(12)}_{\overline{\infty}|4\%} \approx \frac{1}{i^{(12)}} \approx \frac{1}{.03928487738} \approx 25.45508772 \approx 25.45509.$$

(3) Consider an annuity that pays \$640 at the beginning of each four-month period for twenty-one years. The accumulated value at the end of the twenty-one years is

$$3 \times \$640\ddot{s}^{(4)}_{\overline{21}|i} = 3 \times \$640\left[\frac{(1+i)^{21} - 1}{d^{(3)}}\right].$$

Noting that $1 + i = \left(1 + \frac{i^{(4)}}{4}\right)^4$ and that $d^{(3)} = 3\left[1 - (1+i)^{-\frac{1}{3}}\right] = 3\left[1 - \left(1 + \frac{i^{(4)}}{4}\right)^{-1}\right]$, the time-twenty-one value of our annuity may be written as

$$3 \times \$640\left[\frac{\left(\left(1 + \frac{i^{(4)}}{4}\right)^4\right)^{21} - 1}{3\left[1 - \left(1 + \frac{i^{(4)}}{4}\right)^{-\frac{4}{3}}\right]}\right] = \$640\left[\frac{\left(1 + \frac{i^{(4)}}{4}\right)^{84} - 1}{1 - \left(1 + \frac{i^{(4)}}{4}\right)^{-\frac{4}{3}}}\right].$$

(5) perpetuity-due solution: The present value of the perpetuity is equal to \$13,520, the price that Jacon paid for the perpetuity. Therefore, if we denote the waiting time (*in years*) by T, since the annuity has quarterly \$750 payments and $d^{(4)}/4 = 1 - (1+i)^{-\frac{1}{4}}$ is the quarterly discount rate, we have

$$\$13{,}520 = (1+i)^{-T}\frac{\$750}{1 - (1+i)^{-\frac{1}{4}}}.$$

Therefore,

$$(1+i)^T = \frac{\$750/\$13{,}520}{1 - (1+i)^{-\frac{1}{4}}}, \text{ and } T = \ln\left(\frac{\$750/\$13,520}{1 - (1+i)^{-\frac{1}{4}}}\right)\Big/\ln(1+i).$$

perpetuity-immediate solution: Once again, denote the time until the first payment by T. Then, the value one quarter before the first payment is $\$750/(1+i)^{\frac{1}{4}} - 1$, and therefore

$$\$13{,}520 = (1+i)^{-(T-\frac{1}{4})}\left(\$750/(1+i)^{\frac{1}{4}} - 1\right).$$

It follows that

$$(1+i)^{T-\frac{1}{4}} = \frac{750}{13,520[(1+i)^{\frac{1}{4}} - 1]}, \text{ and } T = \frac{1}{4} + \ln\left(\frac{750}{13,520[(1+i)^{\frac{1}{4}} - 1]}\right)\Big/\ln(1+i).$$

(7) According to Fact (4.3.2) and Equation (3.2.4),

$$a^{(m)}_{\overline{n}|i} = \frac{1}{m}a_{\overline{nm}|J} = \frac{1}{m}\left[\frac{1 - (1+J)^{-nm}}{J}\right].$$

But $J = (1+i)^{\frac{1}{m}} - 1$ and $(1+J)^{-nm} = (1+i)^{-n}$. Therefore,

$$a^{(m)}_{\overline{n}|i} = \frac{1}{m}\left[\frac{1 - (1+i)^{-n}}{(1+i)^{\frac{1}{m}}}\right] = \frac{1 - v^n}{m[(1+i)^{\frac{1}{m}} - 1]}.$$

(4.4) Annuities with payments less frequent than each interest period and payments in arithmetic progression

(1) (a) The effective quarterly interest rate is $j = (1+\frac{.06}{12})^3 - 1 \approx 1.5075125 \approx 1.50751$, and using Equation (3.9.5), the balance just after the last payment (*the sixtieth payment*) is $\$100(Is)_{\overline{60}|j} = \$100\left(\frac{s_{\overline{61}|j}-61}{j}\right) \approx \frac{98.91057815-61}{.015075125} \approx \$251{,}477.7035 \approx \$251{,}477.70$.

(b) The monthly effective interest rate is $\frac{6\%}{12} = .5\%$, and the balance just after the last payment is

$$\$100\left(\frac{s_{\overline{180}|.5\%}}{a_{\overline{3}|.5\%}} - \frac{180}{3}\right) \bigg/ .005s_{\overline{3}|.5\%} \approx \$100\left(\frac{290.8187124}{2.970248138} - 60\right) \bigg/ .015075125$$
$$\approx \$251{,}477.7035 \approx \$251{,}477.70.$$

(3) Chapter 3 method: In this method, we need to calculate the interest rate for the three-year payment period; it is $(1.05)^3 - 1 \approx 15.7625\%$. The time 2 value of the annuity (*the value one payment period before the first payment*) is

$$(I_{\$15{,}000,\$4{,}000}a)_{\overline{10}|15.7625\%} = \$15{,}000a_{\overline{10}|15.7625\%} + \frac{\$4{,}000}{.157625}\left(a_{\overline{10}|15.7625\%} - 10(1.157625)^{-10}\right)$$
$$\approx \$138{,}171.8181.$$

We multiply this by $(1.05)^{-2}$ to obtain the time 0 value; it is about \$123,325.91.

Section (4.4) method: The time 2 value is given by the expression

$$\$15{,}000\left(\frac{a_{\overline{30}|5\%}}{s_{\overline{3}|5\%}}\right) + \frac{\$4{,}000}{.05s_{\overline{3}|5\%}}\left(\frac{a_{\overline{30}|5\%}}{s_{\overline{3}|5\%}} - \frac{30}{3}(1.05)^{-30}\right)$$
$$\approx \$73{,}144.09688 + \$25{,}376.68517(4.876273125 - 2.313774487) \approx \$138{,}171.8181.$$

Once again, multiply \$138,171.8181 by $(1.05)^{-2}$ to obtain the time 0 value, namely \$123,325.91.

(4.5) Annuities with payments more frequent than each interest period and payments in arithmetic progression

(1) The annuity symbol $(I^{(4)}\ddot{a})^{(4)}_{\overline{\infty}|}$ gives the value at the time of the first payment of a perpetuity with quarterly payments; the first payment amount is $\frac{1}{16}$ and each payment is $\frac{1}{16}$ more than its predecessor. Thus, the amount of the k-th payment is $\frac{k}{16}$.

According to Equations (4.5.4),

$$\begin{aligned}(I^{(4)}\ddot{a})^{(4)}_{\overline{\infty}|} &= (1.032)^{\frac{1}{4}}(I^{(4)}a)^{(4)}_{\overline{\infty}|} \\ &= (1.032)^{\frac{1}{4}}\left(\frac{1}{i^{(4)}d^{(4)}}\right) \\ &= (1.032)^{\frac{1}{4}}\left(\frac{1}{\left(4[(1.032)^{\frac{1}{4}}-1]\right)\left(4[1-(1.032)^{-\frac{1}{4}}]\right)}\right) \\ &\approx \$1{,}015.858756 \approx \$1{,}015.86.\end{aligned}$$

(3) The value of each year's payments at the beginning of the year is

$$(\$2{,}400 \times 12)\ddot{a}^{(12)}_{\overline{1}|i} + (\$300 \times 144)(I^{(12)}\ddot{a})^{(12)}_{\overline{1}|i} = \$28{,}800\ddot{a}^{(12)}_{\overline{1}|i} + \$43{,}200(I\ddot{a}^{(12)})^{(12)}_{\overline{1}|i}.$$

Therefore, the value of the twenty-year annuity just before the first payment is

$$\left(\$28{,}800\ddot{a}^{(12)}_{\overline{1}|i} + \$43{,}200(I^{(12)}\ddot{a})^{(12)}_{\overline{1}|i}\right)\ddot{a}_{\overline{20}|i}.$$

In order to evaluate this at a 5% annual effective interest rate, we note that

$$i^{(12)} = 12[(1.05)^{\frac{1}{12}} - 1] \approx .0488894854,$$

$$\ddot{a}^{(12)}_{\overline{1}|5\%} = (1.05)^{\frac{1}{12}}\left(\frac{1-(1.05)^{-1}}{i^{(12)}}\right) \approx (1.004074124)\left(\frac{.04769048}{.0488894854}\right) \approx .977982344,$$

and therefore

$$\begin{aligned}(I^{(12)}\ddot{a})^{(12)}_{\overline{1}|5\%} &= (1.05)^{\frac{1}{12}}(I^{(12)}a)^{(12)}_{\overline{1}|5\%})\\ &= (1.05)^{\frac{1}{12}}\left(\frac{\ddot{a}^{(12)}_{\overline{1}|5\%} - 1(1.05)^{-1}}{i^{(12)}}\right)\\ &= (1.004074124)\left(\frac{.977982344 - .952380952}{.0488894854}\right)\\ &\approx .525791881.\end{aligned}$$

Moreover,

$$\ddot{a}_{\overline{20}|.05} \approx 13.08532086.$$

So, with a 5% annual effective interest rate, the value of the annuity just before the first payment is

$$\begin{aligned}&\left(\$28{,}800\ddot{a}^{(12)}_{\overline{1}|5\%} + \$43{,}200(I^{(12)}\ddot{a})^{(12)}_{\overline{1}|5\%}\right)\ddot{a}_{\overline{20}|5\%}\\ &\approx [(\$28{,}800)(.977982344) + \$43{,}200(.525791881)](13.08532086)\\ &\approx \$665{,}782.4435 \approx \$665{,}782.44.\end{aligned}$$

(5) The effective monthly interest rate for the account into which Bob makes his deposits is $\frac{6\%}{12} = .5\% = .005$, and we note that $.005 \times \$1{,}500 = \$7.50$. So, for each \$1,500 in the $i^{(12)} = 6\%$ account, the 5.2% account receives an end-of-month deposit of \$7.50. Since interest is paid out and there are no mid-quarter deposits, if $k \in \{1, 2, \ldots, 64\}$, the balance in the $i^{(12)} = 6\%$ account during the k-th quarter is $\$1{,}500k$. Moreover, if $k \in \{65, 66, \ldots, 80\}$, then the balance in in the $i^{(12)} = 6\%$ account during the k-th quarter is $64 \times \$1{,}500 = \$96{,}000$. Thus, the 5.2% account receives three level end-of-month deposits each quarter; the amount of each of these is $\$7.50k$ if $k \in \{1, 2, \ldots, 64\}$ and $64 \times \$7.50 = \$480$ if $k \in \{64, 65, \ldots, 80\}$. The balance of the 5.2% account at the end of sixteen years (sixty-four quarters) is $(3 \times \$7.50)(Is)^{(3)}_{\overline{64}|j}$ where $j = (1.052)^{\frac{1}{4}} - 1 \approx 1.27539249\%$ is the quarterly effective interest rate; the reason we had to multiply \$7.50 by 3 is that the symbol $(Is)^{(3)}_{\overline{64}|j}$ has payments of $k/3$ in the k-th quarter. It follows that the balance in the 5.2% account at the end of twenty years is $(3 \times \$7.50)(Is)^{(3)}_{\overline{64}|j}(1.052)^4 + (12 \times \$480)s^{(12)}_{\overline{4}|}$; here, the second term has a factor of 12 because $s^{(12)}_{\overline{4}|}$ has level monthly payments of $\frac{1}{12}$. We calculate

$$\begin{aligned}(3 \times \$7.50)(Is)^{(3)}_{\overline{64}|j}(1.052)^4 &= \$22.50\left(\frac{\ddot{s}_{\overline{64}|j} - 64}{3\left[(1+j)^{\frac{1}{3}} - 1\right]}\right)(1.052)^4\\ &\approx \$62{,}516.92084(1.052)^4 \approx \$76{,}570.3352,\end{aligned}$$

and

$$(12 \times \$480)s_{\overline{4}|^{(12)}} \approx \$5{,}760\left(\frac{(1.052)^4 - 1}{12\left[(1.052)^{\frac{1}{12}} - 1\right]}\right) \approx \$25{,}488.25393.$$

Thus, the 5.2% account has a balance of $\$76{,}570.3352 + \$25{,}488.25393 \approx \$102{,}058.59$ at the end of the twenty years.

Now consider the the $i^{(12)} = 6\%$ account. Since all the interest earned by this account is immediately paid out to the 5.2% account, the balance of the $i^{(12)} = 6\%$ account at the end of the twenty years will be the total money deposited to it; that is $64 \times \$1{,}500 = \$96{,}000$. Bob's total accumulation at the end of twenty years is the sum of the accumulations of the two accounts; it is

$$\$102{,}058.59 + \$96{,}000 = \$198{,}058.59.$$

(7) First, we shall determine the monthly deposits paid to the account with quarterly interest rate j. The monthly interest rate on the account to which Darlene makes her deposits is $\frac{6\%}{12} = .5\%$, and for $k \in \{1, 2, \ldots, 40\}$, the balance in this account during the k-th quarter is $\$1{,}000k$. Therefore, if $k \in \{1, 2, \ldots, 40\}$, at the end of each month of the k-th quarter, the deposit to the account with interest rate j is $(.005)(\$1{,}000k) = \$5k$. Note that if $k = 40$, this deposit amount is \$200. Moreover, since Darlene makes no deposits after the beginning of the fortieth quarter, the account with interest rate j gets level end-of-month deposits of \$200, the interest on \$40,000; of course these stop once the $i^{(12)} = 6\%$ is liquidated.

PAYMENT:		5	5	5	10	10	10	...	\$200	\$200	\$200	\$200	...	\$200
TIME:	0	$\frac{1}{12}$	$\frac{2}{12}$	$\frac{3}{12}$	$\frac{4}{12}$	$\frac{5}{12}$	$\frac{6}{12}$	...	$\frac{117}{12}$	$\frac{118}{12}$	$\frac{119}{12}$	$\frac{120}{12}$	...	$\frac{144}{12}$
TIME:	0			$\frac{1}{4}$			$\frac{2}{4}$	...	$\frac{39}{4}$			$\frac{40}{4}$	...	$\frac{48}{4}$

40 quarters, each with 3 level monthly payments

8 quarters, \$200 monthly payments

Deposits to the interest rate j account

The balance in the quarterly-rate-j account at the end of ten years is $(3 \times \$5)(Is)^{(3)}_{\overline{40}|j}$. Bringing this balance forward two years (*eight quarters*) , contributes $(3 \times \$5)(Is)^{(3)}_{\overline{40}|j}(1+j)^8$ to the total balance. Moreover, the monthly level deposits of \$200 during the last eight quarters contribute an additional $(3 \times \$200)s^{(3)}_{\overline{8}|j}$. So, Darlene's accumulation at the end of the twelve years is

$$\$15(Is)^{(3)}_{\overline{40}|j}(1+j)^8 + \$600s^{(3)}_{\overline{8}|j}$$

from the quarterly-rate -j account plus an additional \$40,000 from the account to which she made her forty \$1,000 deposits. Her total liquidated balance is therefore

$$\$15(Is)^{(3)}_{\overline{40}|j}(1+j)^8 + \$600s^{(3)}_{\overline{8}|j} + \$40{,}000.$$

(9) The annual effective interest rate i is equivalent to an effective interest rate $J = \frac{i^{(m)}}{m} = (1+i)^{\frac{1}{m}} - 1$ for the payment period, an m-th of a year. If $k \in \{1, 2, \ldots, n\}$, the payments in the k-th interest period are each for the amount $[P + (k-1)Q]$, and their accumulated value at the end of that interest period is $[P+(k-1)Q]s_{\overline{m}|J} = Ps_{\overline{m}|J} + (Q-1)s_{\overline{m}|J}$. So, the end-of-interest-period accumulated values form an arithmetic progression, the first term of which is $Ps_{\overline{m}|J}$; each term is $Qs_{\overline{m}|J}$ more than its predecessor. If we look at the total accumulated value at the end of n interest periods, it is

$$(I_{Ps_{\overline{m}|J}, Qs_{\overline{m}|J}}s)_{\overline{n}|i} = (Ps_{\overline{m}|J})s_{\overline{n}|i} + \frac{Qs_{\overline{m}|J}}{i}\left(s_{\overline{n}|i} - n\right).$$

Next observe that

$$s_{\overline{m}|J}s_{\overline{n}|i} = ms^{(m)}_{\overline{n}|i};$$

this is because each side of the equation represent the accumulated value of payments of 1 at the end of each m-th of an interest period for n interest periods. It follows that the requested accumulation value, that we last expressed as $(Ps_{\overline{m}|J})s_{\overline{n}|i} + \frac{Qs_{\overline{m}|J}}{i}\left(s_{\overline{n}|i} - n\right)$, may be rewritten as

$$mPs^{(m)}_{\overline{n}|i} + \frac{Qs_{\overline{m}|J}}{i}\left(s_{\overline{n}|i} - n\right) = m\left(Ps^{(m)}_{\overline{n}|i} + \frac{Qs_{\overline{m}|J}}{mi}\left(s_{\overline{n}|i} - n\right)\right).$$

Since $(1+J)^m = (1+i)$,

$$\frac{Qs_{\overline{m}|J}}{mi} = \frac{Q[(1+J)^m - 1]/J}{mi} = \frac{Q}{mJ} = \frac{Q}{i^{(m)}},$$

and therefore our formula for the accumulated value may finally be rewritten as $m\left[Ps^{(m)}_{\overline{n}|i} + \frac{Q}{i^{(m)}}(s_{\overline{n}|i} - n)\right]$.

(4.6) Continuously paying annuities

(1) The annuity symbol $\bar{a}_{\overline{18}|3.2\%}$ represents the present value of an annuity which pays continuously in a level manner for eighteen interest periods, with a total of 1 being paid each interest period; the effective interest rate for the interest period is 3.2%.

$$\bar{a}_{\overline{18}|} = \frac{1 - v^{18}}{\delta} = \frac{1 - (1.032)^{-18}}{\ln 1.032} \approx 13.73905211 \approx 13.739.$$

(3) We apply Fact (4.6.9) with $f(t) = (2+t)^2$ and

$$v(t) = e^{-\int_0^t (1+r)^{-1}\,dr} = e^{-\ln(1+r)|_0^t} = e^{-\ln(1+t)} = e^{\ln(1+t)^{-1}} = (1+t)^{-1} = \frac{1}{1+t}.$$

The present value of the annuity is

$$\int_0^{10} f(t)v(t)\,dt = \int_0^{10} (2+t)^2 \frac{1}{1+t}\,dt = \int_0^{10} \frac{t^2 + 4t + 4}{t+1}\,dt.$$

Therefore, to find the requested present value, we wish to integrate a rational function, the numerator of which has higher degree than the denominator; such a function is called an "improper rational function".

The standard first step for the integrastion of an improper rational function is to divide the polynomial in the numerator by the one in the denominator and use the result to write the rational function as a sum of a polynomial and a rational function; in this case,

$$t^2 + 4t + 4 = (t+1)(t+3) + 1,$$

and

$$\frac{t^2+4t+4}{t+1} = \frac{(t+1)(t+3)+1}{t+1} = t+3+\frac{1}{t+1}.$$

It follows that

$$\int_0^{10} \frac{t^2+4t+4}{t+1}\,dt = \int_0^{10} t+3+\frac{1}{t+1}\,dt = \left[\frac{t^2}{2}+3t+\ln(t+1)\right]\Bigg|_0^{10} \approx 50+30+2.397895273 \approx 82.39790.$$

Assuming the monetary units are dollars, we round to \$82.40.

(5) According to Equation (4.6.8), the answer may be found by calculating $\$72\int_0^{20} v(t)\,dt$. Moreover,

$$v(t) = e^{-\int_0^t \delta_r\,dr} = e^{-\int_0^t \frac{5}{3+2r}\,dr} = e^{-\frac{5}{2}\ln(3+2r)|_0^t}$$
$$= e^{\left[\ln(3+2r)^{-\frac{5}{2}}\Big|_0^t\right]} = e^{\left[\ln\left((3+2t)^{-\frac{5}{2}}\right)-\ln\left(3^{-\frac{5}{2}}\right)\right]}$$
$$= e^{\ln\left(\frac{3+2t}{3}\right)^{-\frac{5}{2}}} = \left(\frac{3+2t}{3}\right)^{-\frac{5}{2}}.$$

Thus, the requested present value is

$$\$72\int_0^{20}\left(\frac{3+2t}{3}\right)^{-\frac{5}{2}}dt = -\$72\left(\frac{3+2t}{3}\right)^{-\frac{3}{2}}\Bigg|_0^{20} \approx \$70.67318096 \approx \$70.67.$$

(7) (a) The interval $[\frac{j-1}{m}, \frac{j}{m}]$ has length $\left(\frac{j}{m}-\frac{j-1}{m}\right) = \frac{1}{m}$. The only payment during the interval is an end-of-interval payment of for an amount $\frac{j}{m^2}$, so the total paid during the interval is $\frac{j}{m^2}$. So, the rate of payment for the interval is $\frac{j}{m^2}/\frac{1}{m} = \frac{j}{m}$.

(b) This annuity has mn payments, and the amount of the k-th payment is $\frac{k}{m^2}$. Therefore, the total amount paid is

$$\sum_{k=1}^{nm}\frac{k}{m^2} = \frac{1}{m^2}\left(\frac{1+2+\cdots+nm}{nm}\right).$$

A useful formula for the sum of the first N integers is $1+2+\cdots+N = \frac{N(N+1)}{2}$; this may be proved by noting that the sum is equal to

$$\frac{1}{2}\left[(1+2+\cdots+N)+(N+(N-1)+\cdots+1)\right] = \frac{1}{2}\left[(1+N)+(2+(N-1))+\cdots+(N+1)\right] = \frac{N(N+1)}{2}.$$

Applying this with $N = nm$, we obtain that the total amount of the payments is

$$\frac{1}{m^2}\left(\frac{nm(nm+1)}{2}\right) = \frac{n^2m^2}{2m^2}+\frac{nm}{2m^2} = \frac{n^2}{2}+\frac{n}{2m}.$$

(c) Note that

$$\lim_{m\to\infty}\left(\frac{n^2}{2}+\frac{n}{2m}\right) = \frac{n^2}{2} = \frac{t^2}{2}\Bigg|_0^n = \int_0^n t\,dt.$$

The limit is the limit of the total payment amounts found in (b). On the other hand, the annuity underlying the symbol $(\overline{I}\overline{a})_{\overline{n}|}$ pays at a rate t at time t; this follows from the result of part (a), since the annuity underlying the symbol $(\overline{I}\overline{a})_{\overline{n}|}$ is the limiting annuity obtained from the annuities underlying the symbols $(I^{(m)}a)^{(m)}_{\overline{n}|i}$ *(as m approaches infinity)*. So, the integral $\int_0^n t\,dt$ is the sum of the total payments of the annuity underlying the symbol $(\overline{I}\overline{a})_{\overline{n}|}$.

(4.7) A yield rate example

(1) Let

$$f(x) = 7{,}200\left(\frac{(1+x)^{15}-1}{4\left[1-(1+x)^{-.25}\right]}\right)(1+x)^5 - 219{,}296.76$$
$$= 1{,}800\left[(1+x)^{20}-(1+x)^5\right]\left[1-(1+x)^{-.25}\right]^{-1} - 219{,}296.76.$$

Then,

$$f'(x) = 1{,}800\left[20(1+x)^{19}-5(1+x)^4\right]\left[1-(1+x)^{-.25}\right]^{-1} +$$
$$(.25)(1,800)\left[(1+x)^{20}-(1+x)^5\right]\left[1-(1+x)^{-.25}\right]^{-2}(1+x)^{-1.25}.$$

We set $x_1 = .052$ and calculate

$$x_2 = x_1 - \frac{f(x_1)}{f'(x_1)} = .052 - \frac{f(.052)}{f'(.052)} \approx .052 - \left(\frac{-9{,}507.842025}{10{,}523,310.16}\right) \approx .052903503,$$

and

$$x_3 = x_2 - \frac{f(x_2)}{f'(x_2)} \approx .052903503 - \left(\frac{-7{,}048.884269}{10{,}508,230.85}\right) \approx .053574299.$$

At this point, it appears that the root is close to 5.4%. In fact, if we calculate the next two estimates given by Newton's method , we find

$$x_4 = x_3 - \frac{f(x_3)}{f'(x_3)} \approx .053574299 - \left(\frac{-5{,}204.16141}{10{,}498,960.1}\right) \approx .054069983,$$

and then

$$x_5 = x_4 - \frac{f(x_4)}{f'(x_4)} \approx .054069983 - \left(\frac{-3{,}830.456486}{10{,}493,132.6}\right) \approx .054435027,$$

confirming that this is correct.

(3) Let $J = (1.054)^{\frac{1}{12}}$; this is the effective monthly rate of interest that is equivalent to the 5.4% annual effective interest rate. The balance in the 5.4% account at the end of the first year is $\$100\ddot{s}_{\overline{12}|J} \approx$ \$1,234.817986. Therefore, since a total of \$1,200 was deposited during the first year, the amount of interest paid out at the end of the year is \$34.82. The deposits in each of the next four years are identical to those in the first year; hence, they again contribute \$34.82 of interest. This is paid out along with interest on the balance that was in place at the beginning of the year; this balance is \$1,200 at the beginning of the second year, $2 \times \$1{,}200 = \$2,400$ at the beginning of the third year, $3 \times \$1{,}200 = \$3,600$ at the beginning of the fourth year, $4 \times \$1{,}200 = \$4,800$ at the beginning of the fifth year, and $5 \times \$1{,}200 = \$6,000$ at the beginning of years six, seven, and eight. Therefore, since $(4.054)(\$1{,}200) = \$64.80$, the interest deposits to the $i^{(4)} = 4\%$ account are as follows:

$$\$34.82 \quad \text{at } t = 1$$
$$\$34.82 + \$64.80 = \$99.62 \text{ at } t = 2,$$
$$\$34.82 + 2(\$64.80) = \$164.42 \text{ at } t = 3,$$
$$\$34.82 + 3(\$64.80) = \$229.22 \text{ at } t = 4,$$
$$\$34.82 + 4(\$64.80) = \$294.02 \text{ at } t = 5, \text{ and}$$
$$5(\$64.80) = \$324 \quad \text{at times } 6, 7, \text{ and } 8.$$

Let $i = \left(1+\frac{.04}{4}\right)^4 - 1 = 4.060401$, the annual effective interest rate for the account that receives the interest. Then, the accumulated value of this account at the the end of eight years is

$$(I_{\$34.82,\$64.80}s)_{\overline{5}|i}(1+i)^3 + \$324s_{\overline{3}|i} \approx \$1{,}985.210808 \approx \$1{,}985.21.$$

[*With only five payments for the annuity with payments in arithmetic progression, you can bring forward the payments one-by-one, or use* Equation (3.9.3).] Combining this with the \$6,000 from the 5.4% account, Daphne has \$7,985.21 at the end of eight years.

Daphne makes deposits of \$100 at the beginning of each month for five years (*60 months*), and she receives a liquidated payment of \$7,985.21 at the end of eight years (*96 months*). The monthly effective interest rate q thus satisfies the time eight 96 months equation of value

$$\$100\ddot{s}_{\overline{60}|q}(1+q)^{36} = \$7{,}985.21.$$

You may then find the rate q by "guess and check" or Newton's method, but it is easier to use the **Cash Flow worksheet**. To do so, set CFo = −100, C01 = −100, F01 = 59, C02 = 0, F02 = 36, C03 = 7,985.21, and F03. Keying **IRR** **CPT** will then give you "IRR = .426644486". So, $q \approx .426644486\%$ and the annual yield rate is $(1.00426644486)^{12} - 1 \approx 5.241595711\% \approx 5.24160\%$.

Note: The interest is paid out each year, so we rounded it to the nearest integer number of cents. If you do not do so, your liquidated amount would be \$7,985.20, and the yield rate would be about 5.24157%.

Chapter 4 review problems

(1) (a) The annuity symbol $\bar{s}_{\overline{30}|}$ gives the accumulated value at the end of the thirty periods of an annuity that pays levelly and continuously for thirty periods; a total of 1 is paid each period and the effective interest rate for each period is 3%. By Equation (4.6.2),

$$\bar{s}_{\overline{30}|} = \frac{(1.03)^{30} - 1}{\ln(1.03)} \approx 48.28553131 \approx 48.286.$$

(b) The annuity symbol $(I^{(12)}a)^{(12)}_{\overline{\infty}|}$ represents the present value of a perpetuity with payments at the end each twelfth of a period; the k-th payment amount is $\frac{k}{144}$; the effective interest rate for each period is 3%. Using Equation (4.5.4) with

$$(I^{(12)}a)^{(12)}_{\overline{\infty}|} = \frac{1}{i^{(12)}d^{(12)}} = \frac{1}{\left(12[1-(1.03)^{-\frac{1}{12}}-1]\right)\left(12[(1.03)^{\frac{1}{12}}-1]\right)} \approx 1{,}144.527196 \approx 1{,}144.527.$$

(c) The annuity symbol $(Ia)^{(4)}_{\overline{\infty}|}$ gives the present value of a perpetuity with payments at the end of each quarter of a period; the payments in the k-th quarter are each for an amount $\frac{k}{4}$. Using Equation (4.5.9) with a 3% effective interest rate for each interest period, we find

$$(Ia)^{(4)}_{\overline{\infty}|} = \frac{1}{di^{(4)}} = \frac{1}{(.03/1.03)\left(4[(1.03)^{\frac{1}{4}}-1]\right)} \approx 1{,}157.24016 \approx 1{,}157.240.$$

(3) The same sequence of payments occurs each calendar year, and it is an arithmetic progression. Each year, the payments form an annuity-due with twelve monthly payments. Moreover, if $J = (1+i)^{\frac{1}{12}} - 1$ is the effective monthly interest rate, the accumulated value of this twelve-payment annuity at the end of the calendar year is

$$(I_{\$1{,}700,\$300}\ddot{s})_{\overline{12}|J}.$$

It follows that the value of the annuity on the first January 1, just before the first payment, is

$$(I_{\$1{,}700,\$300}\ddot{s})_{\overline{12}|J}\, a_{\overline{15}|i}.$$

Using Equation (3.9.3), if $i = 3\%$,

$$\begin{aligned}(I_{\$1{,}700,\$300}\ddot{s})_{\overline{12}|J} &= (1+J)^{\frac{1}{12}}(I_{\$1{,}700,\$300}s)_{\overline{12}|J}\\ &= (1.03)^{\frac{1}{12}}\left[\$1{,}700 s_{\overline{12}|J} + \frac{\$300}{J}\left(s_{\overline{12}|J} - 12\right)\right]\\ &\approx \$40{,}742.91944\end{aligned}$$

The value of the annuity on the first January 1, just before the first payment, is therefore about $\$40{,}742.91944 a_{\overline{15}|3\%} \approx \$486{,}386.3276 \approx \$486{,}386.33$.

The above value was found by looking at annuity symbols evaluated at interest rate J, the monthly effective interest rate. If you prefer, you can find an expression for the present value just using annuity symbols at the annual effective interest rate i. For fifteen years, there are monthly beginning of month payments, each of which is for at least \$1,400. We think of the given annuity as the sum of an annuity with level beginning-of-month payments of \$1,400 and an annuity that pays $\$300k$ at the beginning of the k-th month, $k \in \{1, 2, \ldots, 12\}$; each of these have a term of fifteen years. Since $12 \times \$1{,}400 = \$16{,}800$, the value of the level \$1,400 annuity at the time of the first monthly payment is

$$\$16{,}800\ddot{a}^{(12)}_{\overline{15}|i}.$$

Moreover, $144 \times \$300 = \$43{,}200$. The January 1 value of each calendar year's payments in the annuity with payments in arithmetic progression is $\$43{,}200(I^{(12)}\ddot{a})^{(12)}_{\overline{1}|}$, so the second summand is

$$\$43{,}200(I^{(12)}\ddot{a})^{(12)}_{\overline{1}|}\ddot{a}_{\overline{15}|i}.$$

So, among the possible expressions giving the specified value of the annuity is

$$\$16{,}800\ddot{a}^{(12)}_{\overline{15}|i} + \$43{,}200(I^{(12)}\ddot{a})^{(12)}_{\overline{1}|}\ddot{a}_{\overline{15}|i}.$$

(5) The identity may be established using Equations (4.3.6) and (4.3.7). We observe that

$$\begin{aligned}\frac{1}{s_{\overline{n}|i}^{(m)}}+i^{(m)} &= \frac{i^{(m)}}{(1+i)^n-1}+\frac{i^{(m)}\left[(1+i)^n-1\right]}{(1+i)^n-1}\\ &= \frac{i^{(m)}(1+i)^n}{(1+i)^n-1}\\ &= \frac{i^{(m)}}{1-v^n}\\ &= \frac{1}{a_{\overline{n}|i}^{(m)}}.\end{aligned}$$

The identity stated in this problem generalizes Equation (3.2.18), and we gave two different proofs of that equation on page 120. Once again, it is possible to give a second proof based on thinking about two different ways you might pay back a loan; this time, the relevant loan would be for an amount m with nm level payments occurring, one at the end of each m-th of an interest period.

(7) We imagine that the given annuity is an annuity with annual payments of \$200, and for which the recipient must immediately refund \$100 in each year divisible by the number 3 (*years 3,6,9,12,15,18,21, and 24*). The annuity with level payments of \$200 has accumulated value $\$200s_{\overline{26}|}$ at the time of the last payment, while the accumulated value of the refunded payments is $\$100\left(\frac{s_{\overline{24}|}}{s_{\overline{3}|}}\right)(1+i)^2$. Thus, the annuity described in the statement of this problem has accumulated value

$$\$200s_{\overline{26}|i}-\$100\left(\frac{s_{\overline{24}|}}{s_{\overline{3}|}}\right)(1+i)^2$$

at the time of its last payment. To finish the problem, note that

$$\begin{aligned}\$200s_{\overline{26}|i}-\$100\left(\frac{s_{\overline{24}|}}{s_{\overline{3}|}}\right)(1+i)^2 &= \$200s_{\overline{26}|i}-\$100\left[\frac{s_{\overline{24}|}}{\left[(1+i)^3-1\right]/i}\right](1+i)^2\\ &= \$200s_{\overline{26}|i}-\$100\left[\frac{s_{\overline{24}|}}{\left[(1+i)-(1+i)^{-2}\right]/i}\right]\\ &= \$200s_{\overline{26}|i}-\$100\left[\frac{s_{\overline{24}|}}{\left(\frac{(1+i)-1}{i}\right)+\left(\frac{1-(1+i)^{-2}}{i}\right)}\right]\\ &= \$200s_{\overline{26}|i}-\$100\frac{s_{\overline{24}|i}}{s_{\overline{1}|i}+a_{\overline{2}|i}}.\end{aligned}$$

CHAPTER 5

Loan repayment

(5.2) Amortized loans and amortization schedules

(1) For use in the completion of the table, we determine the effective interest rate for a time period of length 1. Since the interest on \$22,342 for the interval [1,3] is \$1,916.14, the effective interest rate for a period of length two is $I = \frac{\$1{,}916.14}{\$22{,}342} \approx .085764032$, and the equivalent annual effective interest rate is i= $(1+I)^{\frac{1}{2}} - 1 \approx .042000015$. The initial loan balance L must satisfy the equation $L(1+i) - \$8{,}000 = \$22{,}342$; using the numerically found value of i, we conclude $L = \frac{\$30{,}342}{i} \approx \$29{,}119.00149 \approx \$29{,}119.00$, so the initial loan amount was \$29,119.

To complete the row for line 1, note that the interest in the payment is $\$29{,}119 \approx \$1{,}222.998445 \approx \$1{,}223.00$, and hence the principal is $\$8{,}000 - \$1{,}223 = \$6{,}777$.

Moving down one row to the row for year 3, we first note that the balance just after the payment is $\$22{,}234 - \$9{,}908 = \$12{,}434$ less than it was on the line above; thus the principal in the payment is \$12,434. To find the payment amount at time 3, we calculate the sum of the principal and the interest; namely $\$1{,}916.14\$12{,}434 = \$14{,}350.14$.

For the bottom row, note that the balance is \$0 after the payment, and hence the amount of principal must be the previous balance, namely \$9,908. Moreover, the interest is $\$9{,}908i \approx \$416.1361515 \approx \$416.14$, and the time 4 payment amount is $\$416.14 + \$9{,}908 = \$10{,}324.14$. The completed table is therefore as follows:

TIME	PAYMENT	INTEREST	PRINCIPAL	BALANCE
0	—	—	—	*\$29,119.00*
1	\$8,000	*\$1,223*	*\$6,777*	\$22,342
3	*\$14,350.14*	\$1,916.14	*\$12,434*	\$9,908
4	*\$10,324.14*	*\$416.14*	*\$9,908*	\$0

(3) The amount of the k-th payment is $\$400 + \$45(k-1)$. In particular, the amount of the fourteenth payment is \$985, and the fifteenth payment is for \$1,030. Since the equation $\$400 + \$45(k-1) = \$1{,}480$ has solution $k = [\$1{,}480 - \$400]/\$45 + 1 = 25$, the loan is repaid by 25 payments.

The amount of interest in the fourteenth payment is equal to the difference in the outstanding loan balance B_{13} just after the thirteenth payment and the outstanding loan balance B_{14} just after the fourteenth payment. We note that

$$B_{13} = (I_{\$985,\$45}a)_{\overline{25-13}|4\%} = \$985a_{\overline{12}|4\%} + \frac{\$45}{.04}\left(a_{\overline{12}|4\%} - 12(1.04)^{-12}\right) \approx \$11{,}370.44547,$$

and

$$B_{14} = (I_{\$1{,}030,\$45}a)_{\overline{25-14}|4\%} = \$1{,}030a_{\overline{11}|4\%} + \frac{\$45}{.04}\left(a_{\overline{11}|4\%} - 11(1.04)^{-11}\right) \approx \$10{,}840.26328.$$

The amount of principal in the fourteenth payment is $B_{13} - B_{14} \approx \$11{,}370.44547 - \$10{,}840.26328 \approx \$530.1821814 \approx \$530.18$.

(5) Arlen's mortgage is for \$328,000 – \$33,000 = \$295,000. Let $J = (1.058)^{\frac{1}{12}} - 1$, the effective monthly rate of interest for the loan. To calculate the amount of each of the first 179 payments, solve the equation $Pa_{\overline{180}|J} = \$295{,}000$ and then round up to the nearest penny; you obtain $P \approx 2{,}434.146123$, so the payment amount is \$2,434.15. By the retrospective method, the outstanding loan balance at the end of the fifth year, just after the sixtieth payment, is

$$B_{60} = \$295{,}000(1.058)^5 - \$2{,}434.15 s_{\overline{60}|J} \approx \$222{,}748.7931 \approx \$222{,}748.79.$$

Therefore, the amount of principal paid in the first five years is \$295,000 – \$222,748.79 = \$72,251.21. Since the total paid during the first five years is $60 \times \$2,434.15 = \$146{,}049$, the amount of interest paid in this period is \$146,049 – \$72,251.21 = \$73,797.79.

(7) (a) The effective interest rate for an m-th of an interest period is $J = \frac{i^{(m)}}{m} = (1+i)^{\frac{1}{m}} - 1$, and the given loan has payments of P each m-th of an interest period. At time $\frac{k}{m}$, just after the k-th loan payment, $nm-k$ payments remain and, by the prospective method, the loan balance is $Pa_{\overline{nm-k}|J}$. But

$$Pa_{\overline{nm-k}|J} = P\left(\frac{1-(1+J)^{-(nm-k)}}{J}\right) = P\left(\frac{1-\left((1+i)^{\frac{1}{m}}\right)^{-(nm-k)}}{i^{(m)}/m}\right)$$

$$= mP\left(\frac{1-(1+i)^{-(n-k/m)}}{i^{(m)}}\right) = mP\left(\frac{1-v^{(n-k/m)}}{i^{(m)}}\right) = mPa^{(m)}_{\overline{n-k/m}|i}.$$

(b) The interest at time $\frac{k+1}{m}$ is equal to J times the post-payment loan balance at time $\frac{k}{m}$. By part (a), this loan balance is

$$mPa^{(m)}_{\overline{n-k/m}|i} = mP\left(\frac{1-v^{(n-k/m)}}{i^{(m)}}\right) = P\left(\frac{1-v^{(n-k/m)}}{J}\right),$$

so the interest is $J\left[P\left(\frac{1-v^{(n-k/m)}}{J}\right)\right] = P(1-v^{n-\frac{k}{m}})$.

(5.3) The sinking fund method

(1) We first work on the column giving the interest paid on the loan amount. We note that the interest on the loan (LOAN INT) for the one year interval [0,1] is \$889, and hence the interest for any one-year interval is \$889. In particular, the interest in the loan at time 2 is \$889. In order to find the interest on the loan for the two-year interval [2,4], we need to find the effective two-year interest rate I on the loan. To do this, observe that since the loan amount is \$14,000, the one-year effective interest rate is $i = \frac{\$889}{\$14{,}000} = .0635$. Then $I = (1.035)^2 - 1 = .13103225$, and the interest on the loan at time 4 is $\$14{,}000I = \$1{,}834.4515 \approx \$1{,}834.45$.

Next, let us determine the effective interest rate j earned by the sinking fund account. Observe that at time 2, the sinking fund account is credited with \$218.40 interest, based on there having been a balance of \$5,200 for one year. It follows that $j = \frac{218.40}{\$5{,}200} = .042$.

Now we are ready to systematically complete the table. The line for year 1 should have \$5,200 inserted as the sinking fund deposit (SF DEPOSIT), since the sinking fund balance (SF BAL) is always obtained by summing the previous sinking fund balance, the sinking fund deposit, and the interest on the sinking fund (SF INT); in this case the previous sinking fund fund balance and hence the interest on the sinking fund are both 0. The net balance of the loan (NET BAL) is the difference \$14,000 – \$5,200 = \$8,800 between the loan amount and the sinking fund balance. In line 2, we are given a \$3,000 sinking fund deposit and \$218.40 as the sinking fund interest. Moreover, the previous line's sinking fund balance was \$5,200. Thus, the new sinking fund balance is \$5,200 + \$3,000 + \$218.14 = \$8,418.40. Consequently, the net balance is \$14,000 – \$8,418.40 = \$5,581.60.

At time 4, two years have elapsed since interest has been credited. Therefore, the amount of interest on the sinking fund account is $\$8{,}418.40[(1+j)^2 - 1] = \$8{,}418.40[1.042)^2 - 1] \approx \$721.9956576 \approx$ \$722.00. Moreover, the time 4 net balance is 0, so the SF balance must equal the loan amount, hence be \$14,000. Let X denote the amount of the sinking fund deposit at time 4. Then $\$14{,}000 = \$8{,}418.40 + \$722 = X$, and $X = \$4{,}859.60$.

Filling in these values, the table looks as follow.

TIME	LOAN INT	SF DEPOSIT	SF INT	SF BAL	NET BAL
0	0	0	0	0	\$14,000
1	\$889	*\$5,200*	0	\$5,200	*\$8,800*
2	*\$889*	\$3,000	\$218.40	*\$8,418.40*	*\$5,581.60*
4	*\$1,834.45*	*\$4,859.60*	*\$722.00*	*\$14,000*	\$0

(3) The quarterly interest on the loan is $\$18{,}000\left(\frac{.08}{4}\right) = \$360$ for $6 \times 4 = 24$ quarters, then $\$18{,}000\left(\frac{.12}{4}\right) = \$540$ for $2 \times 4 = 8$ quarters.

Quarterly payments occur for eight years, so there are thirty-two payments. Each end of quarter payment is for \$770 and consists of the interest due on the loan and a deposit to the sinking fund. So, there are twenty-four deposits of \$770 – \$360 = \$410, followed by eight deposits of \$770 – \$540 = \$230. Since the sinking fund has an effective quarterly interest rate of $\frac{6\%}{4} = 1.5\%$, the accumulated value of the thirty-two payments (*at the time of the last payment*) is

$$\$410 s_{\overline{24}|1.5\%}(1.015)^8 + \$230 s_{\overline{8}|1.5\%} \approx \$15{,}164.28705 \approx \$15{,}164.29.$$

Thus, the SF account is short by \$18,000 – \$15,164.29 = \$2,835.71.

(5) Denote the amount of each of Bob's twelve end-of-year payments by X. Then X should satisfy the equation $Xa_{\overline{12}|5.62499\%} = \$10{,}000$, and therefore $X \approx \$1{,}168.370178$; Thus, $X = \$1{,}168.37$. On the other hand, if Y denotes the amount of each of Barbara's twelve deposits to her 4% sinking fund account, then Y should satisfy the equation $Ys_{\overline{12}|4\%} = \$10{,}000$. Therefore, $Y \approx \$665.5217269$, and $Y = \$665.52$. We are given that

$$X - Y = \$10{,}000i,$$

so $1{,}168.37 - \$665.52 = \$10{,}000i$. It follows that $i = 5.0285\%$.

(5.4) Loans with other repayment patterns

(1) First note that the effective discount rate for the payment period is $4\%/4 = 1\%$, so the discount factor for the payment period is $1 - .01 = .99$. There are $8 \times 4 = 32$ end-of-period payments, the first of which is for an amount that we represent by X. Then, the k-th payment is for an amount $X(1.02)^{(k-1)}$ and

has present value $X(1.02)^{(k-1)}(.99)^k$. Therefore, since the loan amount is \$39,999.85,

$$\begin{aligned}\$39,999.85 &= \sum_{k=1}^{32} X(1.02)^{(k-1)}(.99)^k \\ &= X(.99)\left(\frac{[(1.02)(.99)]^{32}-1}{(1.02)(.99)-1}\right);\end{aligned}$$

here we have used Equation (3.2.2) to compute the sum of a geometric series. It follows that

$$X = \left(\frac{\$39{,}999.85}{.99}\right)\left(\frac{(1.02)(.99)-1}{[(1.02)(.99)]^{32}-1} - 1\right) \approx \$1{,}081.099917 \approx \$1{,}081.10.$$

We next calculate the outstanding loan balance just after the twentieth payment, at which point there are $32 - 20 = 12$ payments remaining; the next of these should be for $\$X(1.02)^{20}$. (*Note that if you calculate this using $\$X = \$1{,}081.10$, you obtain 1,574.958559, but of course the actual payment must be an integer number of cents. If you choose to round each payment to the nearest cent, you cannot use the formula for the sum of a geometric cent, so we do not round at this point. This is standard procedure.*)
By the prospective method, the outstanding loan balance B_{20} just after the twentieth payment is

$$\begin{aligned}B_{20} &= \sum_{k=21}^{32} X(1.02)^{k-1}.99^{(k-20)} \\ &= (1.02)^{20}(.99)(\$1{,}081.10)\left(\frac{[(1.02)(.99)]^{12}-1}{(1.02)(.99)-1}\right) \approx \$20{,}147.73992. \approx \$20{,}147.74.\end{aligned}$$

The outstanding loan balance B_{19} just after the nineteenth payment may be calculated in a similar manner;

$$\begin{aligned}B_{19} &= \sum_{k=20}^{32} X(1.02)^{k-1}.99^{(k-19)} \\ &= (1.02)^{19}(.99)(\$1{,}081.10)\left(\frac{[(1.02)(.99)]^{13}-1}{(1.02)(.99)-1}\right) \approx \$21{,}505.47149.\end{aligned}$$

The principal in the twentieth payment is $B_{19} - B_{20} \approx \$21{,}505.47149 - \$20{,}147.73992 \approx \$1{,}357.731574 \approx$ \$1,357.73.

The interest rate for the payment period is .01/.99, and the interest in the twentieth payment is $\left(\frac{.01}{.99}\right) B_{19} \approx \left(\frac{.01}{.99}\right)(\$21{,}505.47149) \approx \$217.2269848 \approx \$217.23$.

We note that you might have also found the interest by finding the amount of the twentieth payment to be $X(1.02)^{(20-1)} = \$1{,}081.10(1.02)^{19} \approx \$1{,}574,958559$ and then subtracting \$1,357.731574, the amount of the principal; you would obtain again obtain about \$217.23.

(3) Taking the given interest rate to be an <u>annual</u> effective rate, the interest rate for the payment period is $j = (1.04)^{\frac{1}{12}} - 1 \approx .327373978\%$. The k-th payment amount is $\$320 + \$5(k-1)$, and $\$320 + \$5(k-1) = \$950$ when $k = 127$, so there are 127 payments. Also note that the sixtieth payment is for \$615, and the sixty-first is for \$620. Therefore, the outstanding loan balance just after the fifty-nineth payment is

$$(I_{\$615,\$5}a)_{\overline{127-59}|J} = \$615a_{\overline{68}|J} + \frac{\$5}{J}\left(a_{\overline{68}|J} - 68(1+J)^{-68}\right) \approx \$47{,}250.91889,$$

and the outstanding loan balance just after the sixtieth payment is

$$(I_{\$620,\$5}a)_{\overline{127-60}|J} = \$620a_{\overline{67}|J} + \frac{\$5}{J}\left(a_{\overline{67}|J} - 67(1+J)^{-67}\right) \approx \$46{,}790.6061.$$

Therefore, the amount of principal in the sixtieth payment is \$47,250.91889 − \$46,790.6061 ≈ \$460.3127871 ≈ \$460.31.

(5) Let X denote the amount of Mr. Beltram's first payment. Then,

$$\begin{aligned}\$100{,}000 &= X(1.02)^{-1} + X(1.1)(1.02)^{-2} + X(1.1)^2(1.02)^{-2}(1.06)^{-1} + \quad X(1.1)^3(1.02)^{-2}(1.06)^{-2} + \\ &\quad X(1.1)^4(1.02)^{-2}(1.06)^{-3} + \cdots + X(1.1)^{11}(1.02)^{-2}(1.06)^{-10} \\ &= X\left[(1.02)^{-1} + (1.1)(1.02)^{-2} + S\right]\end{aligned}$$

where

$$\begin{aligned}S &= (1.1)^2(1.02)^{-2}(1.06)^{-1} + (1.1)^3(1.02)^{-2}(1.06)^{-2} + \cdots + (1.1)^{11}(1.02)^{-2}(1.06)^{-10} \\ &= (1.1)^2(1.02)^{-2}(1.06)^{-1}\left(\frac{\left(\frac{1.1}{1.06}\right)^{10} - 1}{\left(\frac{1.1}{1.06}\right) - 1}\right) \\ &= (1.1)^2(1.02)^{-2}\left(\frac{\left(\frac{1.1}{1.06}\right)^{10} - 1}{.04}\right) \\ &\approx 13.03541939.\end{aligned}$$

Therefore,

$$\begin{aligned}\$100{,}000 &\approx X\left[(1.02)^{-1} + (1.1)(1.02)^{-2} + S\right] \\ &\approx X\,[.980392157 + 1,057285659 + 13.03541939] \\ &\approx X(15.07309721),\end{aligned}$$

and $X \approx \$100{,}000/15.07309721 \approx \$6{,}634.336569 \approx \$6{,}634.34$.

By the prospective method, the outstanding loan balance B_5 just after the fifth payment is given by the geometric series

$$B_5 = X(1.1)^5(1.06)^{-1} + X(1.1)^6(1.06)^{-2} + \cdots + X(1.1)^{11}(1.06)^{-7}.$$

Therefore, using $X \approx \$6{,}634.34$,

$$\begin{aligned}B_5 &\approx \$6{,}634.34(1.1)^5(1.06)^{-1}\left(\frac{\left(\frac{1.1}{1.06}\right)^{7} - 1}{\left(\frac{1.1}{1.06}\right) - 1}\right) = \$6{,}634.34(1.1)^5\left(\frac{\left(\frac{1.1}{1.06}\right)^{7} - 1}{.04}\right) \\ &\approx \$79{,}068.74705 \approx \$79{,}068.75.\end{aligned}$$

Please note that the answer given in the back of the text is \$79,068.68. This answer was obtained using the retrospective method instead of the prospective method. The discrepancy is due to the rounding in the value of X. Of course, when the payments are in geometric progression, we usually have slight rounding issues when sums are computed using Equation (3.2.2).

(5.5) Yield rate examples and replacement of capital

(1) (a) We first determine the amount Q of the eighteen end-of-year sinking fund deposits. Since the sinking fund account has an annual effective interest rate of 5% and the loan amount is \$47,000, we seek Q with $\$47{,}000 = Qs_{\overline{18}|5\%}$. Therefore, $Q \approx \$1{,}670.672449$. In order to have complete replacement of capital, we round up the payment amount to be \$1,670.68. Note that $\$1{,}670.68s_{\overline{18}|5\%} \approx \$47{,}000.21$.

To compute a yield rate, we should always take a "bottom line approach"; in this case, we note that

Admiral Capital pays \$47,000 at $t = 0$

Admiral Capital gets a net amount \$4,675 − \$1,670.68 = \$3,004.32 at $t = 1, 2, \ldots, 17$

Admiral Capital gets a net amount \$3,004.32 + \$47,000.21 = \$50,004.53 at $t = 18$.

The yield rate i thus satisfies the time equation of value

$$\$47{,}000 = \$3{,}004.32 a_{\overline{17}|i} + \$50{,}004.53(1+i)^{-18} = \$3{,}004.32 a_{\overline{18}|i} + \$47{,}000.21(1+i)^{-18}.$$

You may now find the yield rate using the "guess and check method", or more easily using the **TVM worksheet**; to use the **TVM worksheet**, make sure you are in END mode with P/Y=C/Y=1, then enter N=18, PV= −47,000, PMT = 3,004.32, FV = 47,000.21, and then key **CPT** **I/Y**. We find $i \approx 6.392184141\% \approx 6.39218\%$. This rate may also be obtained using the **CF worksheet** with CFo = −47,000, C01 = 3,004.32, F01 = 17, C02 = 47,000.21, and F02=1.

(b) If the sinking fund account is held by Admiral Capital, then a list of their net cashflows is as follows:

Admiral Capital pays \$47,000 at $t = 0$

Admiral Capital gets a net amount \$4,675 at $t = 1, 2, \ldots, 18$;

the sinking-fund deposits and sinking-fund liquidation are not considered since they are just payments of Admiral Capital to itself. Thus, the yield rate i satisfies the time 0 equation of value $\$47{,}000 = \$4{,}675 a_{\overline{18}|i}$ and $i \approx 7.007732051\% \approx 7.00773\%$.

(3) We have four yield rates to calculate.

(a) The investment is based on a nominal discount rate $d^{(12)} = 6.8\%$. The equivalent annual effective rate of interest i is $i = \left(1 - \frac{d^{(12)}}{12}\right)^{-12} - 1 = 7.0572333461\% \approx 7.05723\%$. This is the yield rate.

(b) Each month, the bank receives interest payments for $\left(\frac{.054}{12}\right)\$100{,}000 = \$450$ as well as sinking fund deposits for an amount \$679.12 since $\$100{,}000/\left(s_{\overline{120}|.04/12\%}\right) \approx \$679.1180483$. Note that $\$679.12 s_{\overline{120}|.04/12\%} \approx \$100{,}000.2874 \approx \$\$100{,}000.29$, so the last sinking fund deposit is reduced twenty-nine cents. The bank therefore has the following cashflows:

\$100,000 is paid out at $t = 0$;

\$450 + \$679.12 = \$1,129.12 is received at $t = 1, 2, \ldots, 119$ months;

\$450 + \$679.12 − \$.29 = \$1,128.83 is received at $t = 120$ months.

Therefore, the monthly yield rate satisfies the equation of value

$$\$100{,}000 = \$1{,}129.12 a_{\overline{120}|j} - \$.29(1+j)^{-120}$$

and $j \approx .531230668\%$; this is most easily obtained using either the **TVM worksheet** or the **Cashflow worksheet** of the BA II Plus calculator . The monthly yield rate j is equivalent to the annual yield rate $i = (1+j)^{12} - 1 \approx 6.564361912\% \approx 6.56436\%$.

We note that had we not insisted that the sinking fund deposits be an integer number of cents, we would have obtained a yield rate of 6.564357359%, which is again about 6.56436%. This would have been simpler, but we could not guaranty ahead of time that this would have given the same result when the answer is kept to that degree of accuracy.

(c) The loan is made at the effective discount rate $d = 8.2\%$. The yield rate is the equivalent annual effective interest rate $i = .082/(1 - .082) \approx .089324619 \approx 8.93246\%$.

(ALL) For investment (a), the loan has an effective monthly interest rate $I = \left(\frac{.068}{12}\right) / \left[1 - \left(\frac{.068}{12}\right)\right] \approx .569896078\%$, and the monthly payments P are determined by solving the equation $\$120{,}000 = Pa_{\overline{120}|I}$; it follows that $P \approx \$1{,}383.349663 \approx \$1{,}383.35$. So, the first 119 payments are each for \$1,383.35, and the last payment may be determined proceeding as in Algorithm (3.2.14) or Algorithm (3.2.19); the final payment amount is \$1,383.29.

For investment (c), the amount repaid is \$470,547.19; this is because the loan is made at an annual effective discount rate numerically equal to 8.2%, and $\$200{,}000\left((1-.082)^{-10}\right) \approx \$470{,}547.1893$.

The bank's combined cash flows are as follows;

$\$120{,}000 + \$100{,}000 + \$200{,}000 = \$420{,}000$ is paid out at $t = 0$;

$\$1{,}383.35 + \$1{,}129.12 = \$2{,}512.47$ is received at $t = 1, 2, \ldots, 119$ months;

$\$1{,}383.29 + \$1{,}128.83 + \$470{,}547.19 = \$473{,}059.31$ is received at $t = 120$ months.

Therefore, the monthly yield rate q satisfies the equation of value

$$\$420{,}000 = \$2{,}512.47a_{\overline{120}|q} + (\$473{,}059.31 - \$2{,}512.47)(1+q)^{-120},$$

and $q \approx .0664104278$. The equivalent annual yield is $(1+q)^{12} - 1 \approx 8.266875052\% \approx 8.2668\%$.

(5) Sheryl's monthly yield rate J satisfies the equation of value $\$22{,}000 = \$245a_{\overline{120}|J}$. Therefore, $J \approx .505688971\%$. Her annual yield rate is $(1+J)^{12} - 1 \approx 6.239921308\%$. The lender, in order to save an amount \$22,000, makes a level deposit of X at the end of each month to the sinking fund account. The effective monthly interest rate on the sinking fund account is $q = (1.03)^{\frac{1}{12}} - 1 \approx .246626977\%$ during the first forty-eight months, and $r = \frac{.04}{12} / \left(1 - \frac{.04}{12}\right) = \frac{.04}{12-.04} \approx .334448161\%$ for the next forty-eight months. Therefore, we seek X with

$$\$22{,}000 = Xs_{\overline{48}|q}\left(1 - \frac{.04}{12}\right)^{-72} + Xs_{\overline{72}|r}.$$

Thus, the level deposit amount X should satisfy

$$X = \$22{,}000 / \left[Xs_{\overline{48}|q}\left(1 - \frac{.04}{12}\right)^{-72} + s_{\overline{72}|r}\right] \approx \$150.7098169.$$

Since X must be an integer number of cents, the lender makes sinking fund deposits at the end of each month for \$150.71. Since the lender also receives \$245, there is a net inflow of $\$245 - \$150.71 = \$94.29$ at the end of the first 119 months.

To calculate the net inflow at the end of the 120-th month, note that

$$\$150.71s_{\overline{48}|q}\left(1 - \frac{.04}{12}\right)^{-72} + \$150.71s_{\overline{72}|r} \approx \$22{,}000.026272 \approx \$22{,}000.03.$$

Thus, due to liquidation of the sinking fund account immediately following the final \$150.71 deposit, the net inflow at the end of the 120th month is $\$22{,}000.03 + \$245 - \$150.71 = \$22{,}094.32$.

A time 0 equation of value satisfied by the lender's monthly yield rate I is

$$\$22{,}000 = \$94.29a_{\overline{120}|I} + \$22{,}000.03(1+I)^{-120}.$$

Consequently, $I \approx 4.285917806\%$. The equivalent annual yield for the lender is $(1+I)^{12} - 1 \approx 5.266086213\% \approx 5.26609\%$.

The difference between the annual effective interest rate charged to Ms. Tran and the lender's annual yield is about $6.239921308\% - 5.266086213\% \approx .9738351\% \approx .97384\%$.

Chapter 5 review problems

(1) Let X denote the payment amount for the \$50,000 amortized loan. Then, we seek X with

$$\$50{,}000 = Xa_{\overline{12}|6\%}.$$

It follows that $X \approx \$5{,}963.851469$; so the first eleven amortization payments are for \$5,963.86 and we have a slightly reduced final payment.

Denote the amount of the first sinking-fund deposit by Y. Since the sinking fund account pays interest an annual effective rate of 3% and the payments increase by 4% annually, we have

$$\$50{,}000 = Y(1.03)^{11} + Y(1.04)(1.03)^{10} + Y(1.04)^2(1.03)^9 + \cdots + Y(1.04)^{11}(1.03)^0.$$

Therefore, recalling Equation (3.2.2) which gives a formula for the sum of a geometric series,

$$\$50{,}000 = Y(1.03)^{11}\left[\frac{\left(\frac{1.04}{1.03}\right)^{12} - 1}{\left(\frac{1.04}{1.03}\right) - 1}\right] = Y(1.03)^{12}\frac{\left(\frac{1.04}{1.03}\right)^{12} - 1}{.01},$$

and $Y \approx \$2{,}852.719809 \approx \$2{,}852.72$.

Exactly five years after he takes out the \$100,000 loan, Dustin will make his level amortized loan payment for \$5,963.85, his sinking fund deposit for $\$2{,}852.72(1.04)^4 \approx \$3{,}337.28$, and an interest payment (on the \$50,000 sinking fund loan) for $(.05)(\$50{,}000) = \$2,500$; therefore, his total payment will be $\$5{,}963.86 + \$3{,}337.28 + \$2{,}500 = \$11{,}801.14$.

(3) We are told that the last five payments on the 4%, \$10,000 loan are for an amount greater than \$1,100; we denote the amount that each of these payments exceed \$1,100 by E. We then have the time 0 equation of value

$$\$10{,}000 = \$1{,}100a_{\overline{10}|4\%} + Ea_{\overline{5}|4\%}(1.04)^{-5}.$$

We therefore should have

$$\$E = (1.04)^5\left(\$10{,}000 - \$1{,}100a_{\overline{10}|4\%}\right)/a_{\overline{5}|4\%} \approx \$294.6141033,$$

so the final five payments are each for $\$1{,}100 + \$294.61 = \$1{,}394.61$ (*or for* \$294.62 *if you do not want the lender to fall short of the desired* \$10,000).

We now come to the meaning of the phrase "that portion representing principal", what this means here is that if we denote the lender's yield rate on his \$10,000 investment by i, so that the portion representing interest is $\$10{,}000i$, the remainder of the payment is "the portion representing principal". So, for the first five payments, the "portion representing principal" is $\$1{,}100 - \$10{,}000i$; for the final five payments, the "portion representing principal" is $(\$1{,}100 + \$294.61) - \$10{,}000i = (\$1{,}100 - \$10{,}000i) + \$294.61$.

We are given that the lender will replace his capital (the invested \$10,000) by means of the 5% sinking fund, and we now understand that the deposits made to the sinking fund at the end of the first five years are each for $(\$1{,}100 - \$10{,}000i)$, while those at the end of the next five years are each for $(\$1{,}100 - \$10{,}000i) + \$294.61$. Therefore,

$$\$10{,}000 = (\$1{,}100 - \$10{,}000i)s_{\overline{10}|5\%} + \$294.61s_{\overline{5}|5\%}.$$

It follows that

$$i = \frac{1}{10{,}000}\left[1{,}100 - \left(\frac{10{,}000 - 294.61s_{\overline{5}|5\%}}{s_{\overline{10}|5\%}}\right)\right] \approx 4.343802427\% \approx 4.34380\%.$$

(*If you used* \$294.62 *rather than* \$294.61, *you would obtain* 4.343846358%. Again, this would round to 4.34380%.)

(5) We begin by calculating the loan amount. Let $J = (1.048)^{\frac{1}{12}} - 1 \approx .391460763\%$. Then the value of the set of sixty monthly \$674 payments, one month before the first payment, is $\$674a_{\overline{60}|J}$. But the first payment occurs in twelve months. Hence, the value of the set of sixty loan payments at the time of the loan is

$$\$674a_{\overline{60}|J}(1.048)^{-\frac{11}{12}} \approx \$34{,}465.82659.$$

Thus, the loan amount is \$34,465.83, and the interest due at the time of the first payment is

$$(.048)(\$34{,}465.83) \approx \$1{,}654.35984 \approx \$1{,}654.36.$$

The outstanding loan balance just after the first payment is $\$34{,}465.38(1.048) - \$674 = \$35{,}446.18984$.

We now have \$35,446.18984 to be paid off over 59 months, and the monthly interest rate is J. Let $v_J = \frac{1}{1+J} = (1.048)^{-\frac{1}{12}}$. Then, as in Amortization Schedule (5.2.6) with $n = 59$, $Q = \$674$, and v_J replacing v, the interest in the k-th payment remaining [the $(k+1)$-st of our original payments] is

$$\$674\left[1 - {v_J}^{59-k+1}\right] = \$674\left[1 - {v_J}^{60-k}\right].$$

We wish to calculate the total interest in the odd loan payments; that is to say, we want the sum of the interest in the first payment (\$1,654.36) and in each of the even numbered payments, once the first payment is removed from the list. Therefore, the desired total is

$$\begin{aligned}&\$1{,}654.36 + \sum_{k=1}^{29} \$674\left[1 - {v_J}^{60-2k}\right]\\ &= \$1{,}654.36 + (\$674)(29) - \$674\left[{v_J}^{58} + {v_J}^{56} + {v_J}^{56} + \cdots + {v_J}^{2}\right].\end{aligned}$$

The sum ${v_J}^{58} + {v_J}^{56} + {v_J}^{56} + \cdots + {v_J}^{2}$ is geometric, so it may be summed using Equation (3.2.2). Since $J = (1.048)^{\frac{1}{12}} - 1 \approx .391460763\%$ and $v_J = (1.048)^{-\frac{1}{12}}$, we find

$$\begin{aligned}{v_J}^{58} + {v_J}^{56} + {v_J}^{56} + \cdots + {v_J}^{2} &= {v_J}^{58}\left[\frac{\left((1+J)^2\right)^{29} - 1}{(1+J)^2 - 1}\right]\\ &= \frac{1 - {v_J}^{58}}{(1+J)^2 - 1}\\ &\approx 25.84773467.\end{aligned}$$

Therefore, the desired total is $\$1{,}654.36 + (\$674)(29) - \$674[25.84773467] \approx \$3{,}778.986831 \approx \$3{,}778.99$.

(7) Let P denote the amount of each of the payments of the first four payments. Then, the next fourteen payments are for the amount $2P$. We have the time 0 equation of value

$$\$75{,}000 = 2Pa_{\overline{18}|6\%} - Pa_{\overline{4}|6\%}.$$

This is equivalent to the equation

$$P = \frac{\$75{,}000}{2a_{\overline{18}|6\%} - a_{\overline{4}|6\%}},$$

so $P \approx 4{,}123.121612$. Since P must be an integer number of cents, $P = \$4{,}123.12$. The final fourteen payments are each for $2P = \$8{,}246.24$. By the perspective method, the outstanding loan balance immediately after the sixth payment is

$$B_6 = \$8{,}246.24a_{\overline{12}|6\%} \approx \$69{,}135.18925.$$

Again by the perspective method, the outstanding loan balance immediately after the seventh payment is

$$B_7 = \$8{,}246.24a_{\overline{11}|6\%} \approx \$65{,}037.06061.$$

The amount of principal in the seventh payment is

$$B_6 - B_7 = \$69{,}135.18925 - \$65{,}037.06061 = \$4{,}098.128645 \approx \$4{,}098.13.$$

We note that the outstanding balances B_6 and B_7 may also be calculated using the retrospective method. If you use this method, you will obtain slightly different values, but their difference is again about \$4,098.13. *The discrepancy is due to the fact that when we found P, there was rounding.*

(9) The given interest rate on the loan is $i^{(12)} = 4.5\%$, so the monthly rate of interest is $4.5\%/12 = .375\%$. Once the payments commence (*one year after the loan is made*) there are twelve payments per year. Therefore, the fortieth payment occurs in the fourth year in which there are payments. Since payments start at \$100 and increase by \$10 each year, the fortieth payment is for \$130. So, the amount of principal in the fortieth payment is $\$130 - (.375)B_{39}$ where B_{39} denotes the principal in the thirty-nineth payment.

To calculate B_{39} , we use the perspective method. Just after the thirty-nineth payment, there remain $72 - 39 = 33$ monthly payments; the first nine of these are for \$130, and then there are twelve \$140 payments followed by twelve \$150 payments. Thus,

$$B_{39} = \$150a_{\overline{33}|.375\%} - \$10a_{\overline{21}|.375\%} - a_{\overline{9}|.375\%} \approx \$4{,}357.87355,$$

and the amount of principal in the fortieth payment is $\$130 - (.00375)B_{39} \approx \$113.6579742 \approx \$113.66$; here we have viewed the \$130 and \$140 payments as \$150 payments with \$20 or \$10 refunded, respectively.

CHAPTER 6

Bonds

(6.2) Bond alphabet soup and the basic price formula

(1) Note that with the standard notation, $F = \$1{,}000$, $N = 10$, $m = 2$, $n = 20$, $\alpha = 10\%$, $r = \frac{\alpha}{m} = \frac{10\%}{2} = 5\%$, $P = \$880$, and $C = \$1{,}020$. The coupon amount is $Fr = (\$1{,}000)(.05) = \$50$, and the basic price formula [Equation (6.2.2)] tells us

$$\$880 = \$50a_{\overline{20}|j} + \$1{,}020(1+j)^{-20}.$$

The effective coupon rate j is hence equal to about $6.109246038 \approx 6.10925\%$; this is easily obtained using either the **TVM worksheet** or the **Cash Flow worksheet** and **IRR** **CPT**.

(3) We start by noting the "bond alphabet soup" for the original bond:
$F = \$1{,}000 = C$, $N = 8$, $m = 2$, $n = 16$, $\alpha = 6\%$, $r = \frac{\alpha}{m} = \frac{6\%}{2} = 3\%$, $Fr = (\$1{,}000)(.03) = \$30$.
According to the basic price formula [Equation (6.2.2)], if this bond is priced to yield 5% nominal interest convertible semiannually (so $I = 5\%$, $j = \frac{I}{m} = \frac{5\%}{2} = 2.5\%$), then

$$P = \$30a_{\overline{16}|2.5\%} + \$1{,}000(1.025)^{-16} \approx \$1{,}065.275013 \approx \$1{,}065.28.$$

For the replacement bond, we modify the standard "alphabet soup" notation by using primes:
$F' = \$1{,}000 = C'$, $m' = 2$, , $\alpha' = 5.5\%$, $r = \frac{\alpha'}{m'} = \frac{5.5\%}{2} = 2.75\%$, $F'r' = (\$1{,}000)(.0275) = \$27.50$.
We seek the number of coupons n' and the term $N' = \frac{n'}{m'} = \frac{n'}{2}$. The price of the replacement bond should be equal to the price of the bond it is replacing. The replacement bond is again priced to yield 5% nominal interest convertible semiannually ($I' = 5\%$, $j' = \frac{I'}{m'} = \frac{5\%}{2} = 2.5\%$), so we have

$$\$1{,}065.28 = P' = \$27.50a_{\overline{n'}|2.5\%} = \$1{,}000(1.025)^{-n'}.$$

Therefore, $n' \approx \$42.84$. But n' must be an integer, so we take $n' = 43$. Thus, the term of the bond is $N' = \frac{43}{2} = 21.5$ years.

(5) The fourteen-year bond has $N = 14$, $m = 2$, $n = 28$, $F = C = \$3{,}000$, $\alpha = 12\%$, $r = 6\%$, and $Fr = (\$3{,}000)(.06) = \$180$. The ten-year replacement bond has $N = 10$, $m = 2$, $n = 20$, $\alpha = 8\%$, and $r = 4\%$. Both bonds have $I = 6\%$ and $j = 3\%$. Therefore, by the basic price formula, the price of the fourteen-year bond is

$$\$180a_{\overline{28}|3\%} + (1.03)^{-28} \approx \$4{,}688.76974 \approx \$4{,}688.77.$$

This must also be the price of the ten-year bond so, denoting the face (and redemption) value of the ten-year bond by $\mathbf{F}$, we have

$$\$4{,}688.77 = \mathbf{F}(.04)a_{\overline{20}|3\%} + \mathbf{F}(1.03)^{-20}.$$

Thus, $\mathbf{F} \approx \$4{,}081.539924$, and the face value of ten year bond is $\mathbf{F} = \$4{,}081.54$.

(7) With all dollar amount given in March 1, 1995 dollars, Bond A has face value $F^A = \$10{,}000$, redemption amount $C^A = \$10{,}000$, $m^A = 1$ coupons per year, coupon rate $r^A = \alpha^A = 4\%$, and inflation-adjusted yield rate $j^A = I^A = 4\%$. Using real (non-inflation adjusted) dollars, Bond B has face amount $F^B = \$10{,}000$, redemption amount $C^B = \$10{,}000$, $m^B = 1$ coupons per year, coupon rate $r^B = \alpha^B = 7\%$, and yield rate $I^B = j^B$ where

$$\$10{,}000 = \$700a_{\overline{10}|j^B} + \$10{,}000(1 + j^B)^{-10}.$$

If q is the annual rate of inflation, then by Equation (1.13.2), the real yield rate (the yield rate for inflation-adjusted dollars) is

$$\frac{1 + j^B}{1 + r^B} - 1.$$

So, if $r = 2.75\%$, the real yield rate is $\frac{1.07}{1.0275} - 1 \approx 4.136253041\%$. Since $4.136253041 > 4$, Bond B is better. If $r = 2.2\%$, the real yield rate is $\frac{1.07}{1.022} - 1 \approx 4.6966732\%$, and Bond B is even better.

(6.3) The premium-discount formula

(1) With the usual notation, we have $F = \$3{,}000$, $m = 1$, $N = n = 12$, $\alpha = r = 9\%$, $I = j = 9.1\%$, and $C - P = \$57$. Therefore, the coupon amount is $Fr = (\$3{,}000)(.09) = \$270$, and the basic price formula gives us

$$P = C - \$57 = \$270a_{\overline{12}|9.1\%} + (1.091)^{-12}.$$

It follows that

$$C\left(1 - (1.091)^{-12}\right) = \$270a_{\overline{12}|9.1\%} + \$57,$$

and

$$C = \left[\$270a_{\overline{12}|9.1\%} + \$57\right] \Big/ \left(1 - (1.091)^{-12}\right) \approx \$3{,}054.947607 \approx \$3{,}054.95.$$

We conclude that $P = \$3{,}054.95 - \$47 = \$2{,}997.95$.

(3) The amount of the premium is $P - C = \$1,400 - \$1,100 = \$300$. Moreover, the bond has $m = 1$ and $n = N = 10$. Therefore, letting D denote the level amount of Alicia's annual deposits to the 8% account, we have $\$300 = Ds_{\overline{10}|8\%}$. Therefore, $D \approx \$20.7088466$. So, her annual payments are for \$20.71; these accumulate to $\$20.71s_{\overline{10}|8\%} \approx \$300.0167087 \approx \$300.02$, so the final payment could be reduced by two cents.

(6.4) Other pricing formulas for bonds

(1) We have $F = \$1{,}000$, $m = 4$, $\alpha = 8\%$, $r = \frac{\alpha}{m} = \frac{8\%}{4} = 2\%$, and $C = \$957$. The coupon amount is $(Fr = \$1{,}000)(.02) = \$20$. The given yield rate is 12% convertible semiannually, and we need an equivalent quarterly yield rate j; $(1 + j)^2 = 1 + \frac{.12}{2}$ so $j \approx 2.95630141\%$. Let n denote the number of coupons in the bond. We are given $K = \$355.40$

Note that the modified coupon rate is $g = \frac{\$20}{\$957}$. According to Makeham's Formula [Equation (6.4.4)],

$$P = \frac{g}{j}(C - K) + K = \frac{.02/957}{(1.06)^{\frac{1}{2}} - 1}(\$1{,}000 - \$355.40) + \$355.40 \approx \$760.6821697 \approx \$760.68.$$

To check the price using another pricing formula, we need to compute n. Since

$$\$355.40 = K = C(1 + J)^{-n} = \$957(1 + j)^{-n} = \$957(1.06)^{-\frac{n}{2}},$$

we find

$$n = 2\ln\left(\frac{957}{355.40}\right) \Big/ \ln(1.06) \approx 33.99958941 \approx 34.$$

Using $n = 34$, the basic price formula gives

$$\begin{aligned} P &= \$20a_{\overline{34}|j} + \$957(1+j)^{-34} \\ &\approx \$425.2851751 + \$355.3957486 \\ &= \$780.6809237 \approx \$760.68. \end{aligned}$$

In each case, we found the price to be \$760.68; the slight discrepancy that was evident before we rounded the price to the nearest cent was due to the fact that we insisted on using an integer value for n, hence used 34 rather than 33.99958941.

(3) Lucia's and Elena's bonds each have $F = C = \$2{,}000$, $m = 4$ $\alpha = 9\%$, $r = g = \frac{\alpha}{4} = 2.25\%$. Elena's bond has $4n$ coupons while Elena's bond has $8n$ coupons; note, the use of n in this problem is different from the use in the standard "bond alphabet soup". The premium-discount formula [Equation (6.3.2)] tells us that the price of Lucia's bond to yield a nominal rate of 6% convertible quarterly (so $I = 6\%$ and $j = 1.5\%$) is

$$P^{\text{Lucia}} = \$2{,}000(.0225 - .015)a_{\overline{4n}|j} + \$2{,}000,$$

and the price of Elena's bond to again yield a nominal rate of 6% convertible quarterly is

$$P^{\text{Elena}} = \$2{,}000(.0225 - .015)a_{\overline{4n}|j} + \$2{,}000.$$

We are given that $P^{\text{Elena}} - P^{\text{Lucia}} = \$233.02$. Therefore,

$$\begin{aligned} \$233.02 &= \left[\$2{,}000(.0225 - .015)a_{\overline{4n}|j} + \$2{,}000.\right] - \left[\$2{,}000(.0225 - .015)a_{\overline{8n}|j} + \$2{,}000.\right] \\ &= \$15a_{\overline{4n}|j} - \$15a_{\overline{8n}|j} \\ &= \$15\left[(1.015)^{-4n} - (1.015)^{-8n}/.015\right] \\ &= \$1{,}000\left[(1.015)^{-4n} - (1.015)^{-8n}\right]. \end{aligned}$$

Setting $x = (1.015)^{-4n}$, we have $233.02 = 1{,}000(x - x^2)$. We may rewrite this as $x^2 - x - .23302 = 0$, and the quadratic equation gives

$$(1.015)^{-4n} = x = \frac{1 \pm \sqrt{1 - 4(.23302)}}{2}.$$

It follows that

$$(1.015)^{-4n} = \frac{1 + \sqrt{1 - 4(.23302)}}{2} \approx .630307329 \quad \text{or} \quad (1.015)^{-4n} = \frac{1 - \sqrt{1 - 4(.23302)}}{2} \approx .369692671.$$

Therefore,

$$4n = \ln(.630307329)/\ln(1.015) \approx 31.00005154 \quad \text{or} \quad 4n = \ln(.369692671)/\ln(1.015) \approx 66.83518951.$$

Since the number of coupons $4n$ in Lucia's bond is given to be an integer, we must have the first of these possibilities, $4n = 31$.

In case you are concerned about our rounding to obtain $4n = 31$, note that if $4n = 31$, then

$$P^{Lucia} = \$15a_{\overline{31}|j} + \$2{,}000 \approx \$2{,}369.692187 \approx \$2{,}369.69,$$

$$P^{Elena} = \$15a_{\overline{62}|j} + \$2{,}000 \approx \$2{,}602.712061 \approx \$2{,}602.71,$$

and $P^{Elena} - P^{Lucia} = \$2{,}602.71 - \$2{,}369.69 \approx \$233.02$ *as desired. On the other hand, if you try rounding* $4n = 66.83518951$ *to* $4n = 67$, *you will find*

$$P^{Lucia} = \$15a_{\overline{67}|j} + \$2{,}000 \approx \$2{,}631.213369 \approx \$2{,}631.21,$$

$$P^{Elena} = \$15a_{\overline{134}|j} + \$2{,}000 \approx \$2{,}863.996421 \approx \$2{,}864.00,$$

and $P^{Elena} - P^{Lucia} = \$2{,}864.00 - \$2{,}631.21 = \$232.79$; *this is a few cents off from the given \$233.02 price difference.*

(6.5) Bond amortization schedules

(1) We are given $F = \$2{,}000$, $m = 2$, $N = 15$, $n = Nm = 30$, $C = \$2{,}100$, $\alpha = 6.5\%$, $r = \frac{\alpha}{m} = \frac{6.5\%}{2} = 3.25\%$, $I = 8\%$, and $j = \frac{8\%}{2} = 4\%$. The coupon amount is $Fr = (\$2{,}000)(.0325) = \$65$. According to the basic price formula,

$$P = \$65a_{\overline{30}|4\%} + \$2{,}100(1.04)^{-30} \approx \$1{,}771.45136 \approx \$1{,}771.45.$$

By the perspective method, again using the basic price formula, the outstanding loan balance at the end of the ninth payment is

$$B_9 = \$65a_{\overline{21}|4\%} + \$2{,}100(1.04)^{-21} \approx \$1{,}833.44596,$$

and the outstanding loan balance at the end of the tenth payment is

$$B_{10} = \$65a_{\overline{20}|4\%} + \$2{,}100(1.04)^{-20} \approx \$1{,}841.783799.$$

Therefore, $P_{10} = B_9 - B_{10} \approx \$1{,}833.44596 - \$1{,}841.783799 \approx -\$8.33783844$, and the amount for accumulation of discount in the tenth payment is $-P_{10} \approx \$8.33783844 \approx \$8.34$. Since the coupon amount is \$65, the amount of interest in the tenth coupon is $I_{10} = \$65 - P_{10} \approx \$65 - (-\$8.33783844) \approx \$73.33783844 \approx \$73.34$.

(3) We are given $m = 2$ and $N = 15$, so the bond has $n = Nm = 30$ coupons. We also know by Equation (6.5.7) that

$$\$977.19 = P_2 = C(g - j){v_j}^{30-2+1} = C(g - j){v_j}^{29},$$

and

$$\$1{,}046.79 = P_4 = C(g - j){v_j}^{30-4+1} = C(g - j){v_j}^{27}.$$

It follows that

$$\frac{1{,}046.79}{977.19} = \frac{C(g - j){v_j}^{27}}{C(g - j){v_j}^{29}} = {v_j}^{-2} = (1 + j)^2.$$

Therefore, $j = \left(\frac{1{,}046.79}{977.19}\right)^{\frac{1}{2}} - 1 \approx 3.49982314$, and

$$C(g - j) = 977.19(1 + j)^{29} = \$2{,}650.006907.$$

By the premium discount formula, the amount of premium in the bond is

$$P - C = C(g - j)a_{\overline{30}|j} = \$2{,}650.006907a_{\overline{30}|j} \approx \$48{,}739.15537 \approx \$48{,}739.16.$$

(5) First note that $N = 3$, $m = 2$, and there are $n = 2 \times 3 = 6$ coupons. Moreover, $F = \$1{,}000$, $C = \$1{,}040$, $\alpha = 6\%$, $r = \frac{6\%}{2}$, and the coupon amount is $Fr = \$30$. We consider the three nominal yield rates convertible semi-annually $I_{[1]} = 5\%$, $I_{[2]} = 6\%$, and $I_{[3]} = 7\%$; these correspond to effective six-month interest rates $j_{[1]} = \frac{I_{[1]}\%}{2} = 2{,}5\%$, $j_{[2]} = \frac{I_{[2]}\%}{2} = 3\%$, and $j_{[3]} = \frac{I_{[3]}\%}{2} = 3.5\%$ for the six-month coupon period. The basic price formula tells us that the prices of the bonds at these three yield rates are

$$P_1 = \$30a_{\overline{6}|2.5\%} + \$1{,}040(1.025)^{-6} \approx \$1{,}062.032501 \approx \$1{,}062.03,$$

$$P_2 = \$30a_{\overline{6}|3\%} + \$1{,}040(1.03)^{-6} \approx \$1{,}033.49937 \approx \$1{,}033.50,$$

and

$$P_3 = \$30a_{\overline{6}|3.5\%} + \$1{,}040(1.035)^{-6} \approx \$1{,}005.897261 \approx \$1{,}005.90,$$

respectively.

Of course, these rounded prices do not give precisely the stated nominal yield rates but they are very close; they correspond to nominal yield rates $I'_{[1]} = 5.0000862254$, $I'_{[2]} = 5.999977553$, *and* $I'_{[3]} = 6.999899098$. *If you use the stated rates rather than* $I'_{[1]}$, $I'_{[2]}$, *and* $I'_{[3]}$, *you may get slightly different results from those we report.*

Our method is to compute I_t from the stored value of B_t using $I_t = jB_{t-1}$ [Equation (6.5.6)], to obtain P_t from I_t using $P_t = \$30 - I_t$ [see Equation (6.5.8)], and to calculate B_t using $B_t = B_{t-1} - P_t$ [line 1 of Equation (6.5.7)]; all reported values are rounded. For example, the entries of the first two row of the "$I_1 = 5\%$" table are obtained as follows:

$$I_1 = 1{,}062.03\left(\frac{I'_{[1]}}{2}\right) \approx \$26.55120787 \approx \$25.56\ ,$$
$$P_1 \approx \$30 - \$26.55120787 \approx \$3.448792132 \approx \$3.45,$$
$$B_1 = B_0 - P_1 = 1{,}062.03 - \$3.448792132 \approx \$1{,}058.581208 \approx \$1{,}058.58,$$
$$I_2 \approx \$1{,}058.581208\left(\frac{I'_{[1]}}{2}\right) \approx \$26.46498658 \approx \$26.46,$$
$$P_2 \approx \$30 - \$26.46498658 \approx \$3.535013422 \approx \$3.54,$$
$$B_2 = B_1 - P_2 = \$1{,}058.581208 - \$3.535013422, \approx \$1{,}055.046194 \approx \$1{,}055.05.$$

The completed tables appear in Appendix B of the text; see page 476.

(6.6) Valuing a bond after its date of issue

(1) The bond has $F = \$2{,}500$, $m = 1$, $n = N = 6$, $r = \alpha = 14\%$, $j = i = 6\%$, $P = \$3{,}432.26$, and the coupon amount is $Fr = (\$2{,}500)(.14) \approx \$350$. Note that $\$3{,}432.26 = P = \$350a_{\overline{6}|6\%} + C(1.06)^{-6}$; thus $C \approx \$2{,}427.36492$, and the redemption amount is $C = \$2{,}427.36$.

We are asked to find the bond's clean and dirty values at the end of each quarter of the fourth year after issue (*by the practical method and also by the theoretical method*); this is at times $3\frac{1}{4}$, $3\frac{2}{4}$, $3\frac{3}{4}$, and 4. We calculate

$$B_3 = \$350a_{\overline{3}|6\%} + \$2{,}427.36(1.06)^{-6} \approx \$2{,}973.612445 \approx \$2{,}973.61,$$

and

$$B_4 = \$350a_{\overline{2}|6\%} + \$2{,}427.36(1.06)^{-6} \approx \$2{,}802.029192 \approx \$2{,}802.03.$$

Therefore, we have theoretical dirty values

$$\begin{aligned}
\mathcal{D}_{3\frac{1}{4}} &= B_3(1.06)^{\frac{1}{4}} \approx \$3{,}017.246777 \approx \$3{,}017.25,\\
\mathcal{D}_{3\frac{2}{4}} &= B_3(1.06)^{\frac{2}{4}} \approx \$3{,}061.521392 \approx \$3{,}061.52,\\
\mathcal{D}_{3\frac{3}{4}} &= B_3(1.06)^{\frac{3}{4}} \approx \$3{,}106.445686 \approx \$3{,}106.45,\\
\mathcal{D}_4 &= B_4 \approx \$2{,}802.03.
\end{aligned}$$

It follows that the theoretical clean values are

$$\begin{aligned}
\mathcal{C}_{3\frac{1}{4}} &= D_{3\frac{1}{4}} - \$350\left(\left[(1.06)^{\frac{1}{4}} - 1\right]/.06\right) \approx \$3{,}017.246777 - \$85.59743599 \approx \$2{,}931.65,\\
\mathcal{C}_{3\frac{2}{4}} &= D_{3\frac{2}{4}} - \$350\left(\left[(1.06)^{\frac{2}{4}} - 1\right]/.06\right) \approx \$3{,}061.521392, -\$172.4509156 \approx \$2{,}889.07,\\
\mathcal{C}_{3\frac{3}{4}} &= D_{3\frac{3}{4}} - \$350\left(\left[(1.06)^{\frac{3}{4}} - 1\right]/.06\right) \approx \$3{,}106.445686 - \$260.5788698 \approx \$2{,}845.87,\\
\mathcal{C}_4 &= B_4 \approx \$2{,}802.03.
\end{aligned}$$

The practical dirty values are

$$\begin{aligned}
\mathcal{D}^{\text{prac}}_{3\frac{1}{4}} &= B_3 - \left[1 + \frac{1}{4}(.06)\right] \approx \$3{,}018.216632 \approx \$3{,}018.22,\\
\mathcal{D}^{\text{prac}}_{3\frac{2}{4}} &= B_3 - \left[1 + \frac{2}{4}(.06)\right] \approx \$3{,}062.820819 \approx \$3{,}062.82,\\
\mathcal{D}^{\text{prac}}_{3\frac{3}{4}} &= B_3 - \left[1 + \frac{3}{4}(.06)\right] \approx \$3{,}107.425005 \approx \$3{,}107.43,\\
\mathcal{D}^{\text{prac}}_4 &= B_4 \approx \$2{,}802.03.
\end{aligned}$$

Lastly, we note that the practical clean values are

$$\begin{aligned}
\mathcal{C}^{\text{prac}}_{3\frac{1}{4}} &= \mathcal{D}^{\text{prac}}_{3\frac{1}{4}} - \frac{1}{4}(\$350) \approx \$3{,}018.216632 - \$87.50 \approx \$2{,}930.72,\\
\mathcal{C}^{\text{prac}}_{3\frac{2}{4}} &= \mathcal{D}^{\text{prac}}_{3\frac{2}{4}} - \frac{2}{4}(\$350) \approx \$3{,}062.820819 - \$175.00 \approx \$2{,}887.82,\\
\mathcal{C}^{\text{prac}}_{3\frac{3}{4}} &= \mathcal{D}^{\text{prac}}_{3\frac{3}{4}} - \frac{3}{4}(\$350) \approx \$3{,}107.425005 - \$262.50 \approx \$2{,}844.93,\\
\mathcal{C}^{\text{prac}}_4 &= B_4 \approx \$2{,}802.03.
\end{aligned}$$

(3) As in Problem(6.5.5), $N = 3$, $m = 2$, and there are $n = 2 \times 3 = 6$ coupons. Moreover, $F = \$1{,}000$, $C = \$1{,}040$, $\alpha = 6\%$, $r = \frac{6\%}{2}$, and the coupon amount is $Fr = \$30$. We are given that the bond was purchased on January 1, 2000 and that it was redeemed on January 1, 2003, and we let the coupon period (*a half-year*) be the basic unit of time; we refer to January 1, 2000 as time 0 and January 1, 2003 as time 6. We are asked for the theoretical and practical dirty and clean values on the bond at the end of each quarter; since we are instructed to use a "30/360" basis for counting days and since a quarter is half of a half year (*our basic unit of time in this solution*), this is at times $\frac{1}{2}, \frac{2}{2}, \ldots, \frac{12}{2}$.

Note that if we write $\frac{k}{2} = \lfloor \frac{k}{2} \rfloor + f_k$ with $0 \le f_k < 1$, then $f_k = 0$ if k is even and $f_k = .5$ if k is odd. Therefore, if k is even, the practical or theoretical clean value is equal to the corresponding dirty value, which in turn is equal to the outstanding loan balance. On the other hand, if k is odd, in general we get four different values; using $j = \frac{7\%}{2} = 3.5\%$, for odd k we have

$$\mathcal{D}_{\frac{k}{2}} = (1.035)^{\frac{1}{2}} B_{\lfloor \frac{k}{2} \rfloor}, \qquad \mathcal{C}_{\frac{k}{2}} = \mathcal{D}_{\frac{k}{2}} - \$30\left(\frac{(1.035)^{\frac{1}{2}} - 1}{.035}\right),$$

$$\mathcal{D}^{\text{prac}}_{\frac{k}{2}} = (1 + (.035)(.5))B_{\lfloor \frac{k}{2} \rfloor}, \quad \text{and} \quad \mathcal{C}^{\text{prac}}_{\frac{k}{2}} = \mathcal{D}^{\text{prac}}_{\frac{k}{2}} - \$15.$$

A complete list of the values obtained by the above formulas is included in Appendix B of the text; see page 477. To make clear how the values are obtained, we give the detailed calculations for the even integer $k = 4$ and for the odd integer $k = 5$. For $k = 4$, we have

$$\mathcal{D}_{\frac{4}{2}} = \mathcal{D}^{\text{prac}}_{\frac{4}{2}} = \mathcal{C}_{\frac{4}{2}} = \mathcal{C}^{\text{prac}}_{\frac{4}{2}} = B_2 = \$30a_{\overline{4}|3.5\%} + \$1{,}040(1.035)^{-4} \approx \$1{,}016.492293 \approx \$1{,}016.49.$$

On the other hand, for the odd integer $k = 5$, we have

$$\mathcal{D}_{\frac{5}{2}} = \mathcal{D}_{\frac{4}{2}}(1.035)^{\frac{1}{2}} \approx (\$1{,}016.492293)(1.017349497) \approx \$1{,}034.127924 \approx \$1{,}034.13,$$

$$\mathcal{C}_{\frac{5}{2}} = \mathcal{D}_{\frac{5}{2}} - \$30\left(\frac{(1.035)^{\frac{1}{2}} - 1}{.035}\right) \approx \$1{,}034.127924 - 14.87099783 \approx \$1{,}019.256926 \approx \$1{,}019.26,$$

$$\mathcal{D}^{\text{prac}}_{\frac{5}{2}} = (1 + (.035)(.5))B_2 \approx (1.0175)(\$1{,}016.492293) \approx \$1{,}034.280908 \approx \$1{,}034.28,$$

and

$$\mathcal{C}^{\text{prac}}_{\frac{5}{2}} = \$1{,}034.280908 - \$15 = \$1{,}019.280908 \approx \$1{,}019.28.$$

(6.7) Selling a bond after its date of issue

(1) The bond has $F = C = \$22{,}000$, $m = 1$, $n = N = 15$, $r = \alpha = 9\%$, $j = i = 7\%$, and coupon amount $Fr = \$1{,}080$. By the basic price formula, Miguel's price is

$$P = \$1{,}080a_{\overline{15}|7\%} + \$22{,}000(1.07)^{-15} \approx \$26{,}007.48216,$$

so he paid \$26,007.48 for the bond. Again by the basic price formula,

$$B_5 = \$1{,}080a_{\overline{15-5}|7\%} + \$22{,}000(1.07)^{-(15-5)} \approx \$25{,}090.37588 \approx \$25{,}090.38.$$

Miguel resells the bond five years after issue (*just after the coupon was paid*) at a price to yield the new buyer $\tilde{j} = 8\%$. Therefore, since ten coupons remain at the the end of five years, the invoice price (*for this resale*) $\mathcal{D}^{\tilde{j}}_T$ should satisfy the equation

$$\mathcal{D}^{\tilde{j}}_T = \$1{,}080a_{\overline{10}|7\%} + \$22{,}000(1.07)^{-10}.$$

It follows that $\mathcal{D}^{\tilde{j}}_T \approx \$23{,}476.21719$, so the invoice price is \$23,476.22, and the difference between Miguel's book value B_5 is \$25,090.38 – \$23,476.22 = \$1,614.16. Miguel's annual yield rate y for the five-year period he held the bond must satisfy the time 0 equation of value

$$\$26{,}007.48 = \$1{,}980a_{\overline{5}|y} + \$23{,}476.22(1 + y)^{-5}.$$

Thus, $y \approx 5.882575186\% \approx 5.88258\%$; this is easily obtained using the **TVM worksheet**.

(3) We have $F = \$18{,}000$, $m = 1$, $r = \alpha = 10\%$, $n = N = 15$, $C = \$19{,}000$, and coupon amount $\$(18{,}000)(.1) = \$1{,}800$.

The number of days from May 27, 2000 to December 31, 2000, figured on a "30/360" basis, is $(31-27)+(12-5)(30) = 214$. Since the "30/360" method calls for a year as being reckoned as 360 days, $f = \frac{214}{360}$, and the accrued interest by the practical method is $(Fr)f = (\$1{,}800)\left(\frac{214}{360}\right) = \$1{,}070$.

We are given that the semipractical clean price is \$18,375. But the semipractical clean price is the difference between the theoretical dirty price and the accrued interest by the practical method. Thus, the theoretical dirty price is $\$18{,}375 + \$1{,}070 = \$19{,}445$; this is the invoice price paid by Edna on December 31, 2000.

Edna's December 31, 2000 payment gave her the right to receive the remaining nine coupons (to be paid on each May 27th through 2009) and also to receive the May 27, 2009 \$18,000 redemption payment. We now find her yield rate assuming she holds the bond through its redemption. We have been instructed to use the "30/360 method. Note that according to this method, there are $(26-1)+(5-1)(30) = 145$ days from January 1, 2001 to May 26, 2001; these are days on which Edna does not receive any payment. Then, there is 1 day (*May 27, 2001*) on which she gets \$1,800. This is followed by 359 days (*a year minus one day*) on which she receives no payment. The same pattern repeats each May 27th through the following May 26th, and then there is a final $19{,}000+1{,}800 = 20{,}800$ payment on May 27, 2009.

If you have the BA II Plus calculator , find Edna's yield rate by using the **Cash Flow worksheet** to find the yield for a $\frac{1}{360}$-th of a year; Use

CF0 = −19,445, (*December 31, 2000 payment*)

C01 = 0, F01 = the number of days (on a "30/360" basis) from January 1, 2001 to May 26, 2001 $= (26-1)+(5-1)(30) = 145$,

C02 = 1,800, F02=1, (*May 27, 2001 payment*)

C03 = C05 = C07= C09 = C11 = C13 = C15 = C17 = 0,

F03=359, F05=359, F07=359, F09=359, F11=359, F13=359, F15=359, F17=359,

C04 = C06 = C08 = C10 = C12 = C14 = C16 =1,800, (*May 27 payments in the seven years 2002, 2003 , 2004, 2005, 2006, 2007, and 2008*)

F04 = F06 = F08 = F10 = F12 = F14 = F16 =1,

C18 = 19,000 + 1,800 = 20,800.

Keying **IRR** **CPT** gives IRR = .026609156. The yearly rate is

$$\jmath = (1.00026609156)^{360} - 1 \approx 10.0517164\% \approx 10.05172\%.$$

If you are not able to use the BA II Plus calculator you will have to use an equation of value to find the yield rate., along with the "guess and check" or "Newton's method". Be prepared for this to take quite a bit of time. A time 0 equation of value that may be used is

$$19{,}445 = \sum_{k=0}^{k=8} \frac{\$1{,}800}{(1+i)^{-\left(\frac{146}{214}+k\right)}}\ .$$

Again, you should find a yield rate $\jmath \approx 10.0517164\%$.

Starting with $\jmath \approx 10.0517164\%$., the accrued coupon by the theoretical method is

$$\$1{,}800\left((1+\jmath)^{\frac{214}{360}} - 1\right) \approx \$1{,}049.159536 \approx \$1{,}049.16.$$

It follows that we have theoretical clean price $C_T^{\jmath} \approx \$19,445 - \$1{,}049.16. \approx \$18{,}395.84$.

(5) Write $T = \lfloor T \rfloor + f$ with $0 \le f < 1$. Let $\widetilde{\jmath} > 0$. Then,

$$\mathcal{A}_T^{\widetilde{\jmath}} = Cgs_{\overline{f}|\widetilde{\jmath}} = Cg\left(\frac{(1+\widetilde{\jmath})^f - 1}{\widetilde{\jmath}}\right) \quad \text{and} \quad \mathcal{A}_T^{\text{prac}} = Cgf.$$

We wish to maximize their difference

$$G(f) = \left|\mathcal{A}_T^{\text{prac}} - \mathcal{A}_T^{\widetilde{\jmath}}\right| = Cg\left|f - \frac{(1+\widetilde{\jmath})^f - 1}{\widetilde{\jmath}}\right|.$$

Since $0 \le f < 1$, simple interest at annual rate $\widetilde{\jmath}$ gives a larger accumulation on $[0, f]$ than compound interest at annual effective rate $\widetilde{\jmath}$; that is, $1 + \widetilde{\jmath}f > (1+\widetilde{\jmath})^f$. Therefore,

$$f = \frac{\widetilde{\jmath}f}{\widetilde{\jmath}} = \frac{(1+\widetilde{\jmath}f) - 1}{\widetilde{\jmath}} > \frac{(1+\widetilde{\jmath})^f - 1}{\widetilde{\jmath}},$$

and thus

$$G(f) = Cg\left(f - \frac{(1+\widetilde{\jmath})^f - 1}{\widetilde{\jmath}}\right).$$

Next consider the derivative of the function $G(f)$ with respect to the variable f. Since we are taking C, g, and $\widetilde{\jmath}$ to be constants, we have

$$G'(f) = Cg\left(1 - \frac{(1+\widetilde{\jmath})^f \ln(1+\widetilde{\jmath})}{\widetilde{\jmath}}\right).$$

Therefore, the condition $G'(f) = 0$ is equivalent with $\widetilde{\jmath} = (1+\widetilde{\jmath})^f \ln(1+\widetilde{\jmath})$ and with

$$(*) \qquad f = \frac{\ln\left[\widetilde{\jmath}/\ln(1+\widetilde{\jmath})\right]}{\ln(1+\widetilde{\jmath})}.$$

Since the second derivative of the function $G(f)$ with respect to the variable f is

$$G''(f) = Cg\left(\frac{-\ln(1+\widetilde{\jmath})}{\widetilde{\jmath}}\right)(1+\widetilde{\jmath})^f \ln(1+\widetilde{\jmath})$$

which is negative for $\widetilde{\jmath} > 0$, Equation (*) gives a maximum.

Let $\widetilde{\jmath} > 0$. We still need to show that the critical value of $G(f)$ given by Equation (*) lies in the interval $[0,1)$. First, we will show that it is positive. We begin by noting that since $\widetilde{\jmath} > 0$ and the natural logarithm is an increasing function, $\ln(1+\widetilde{\jmath}) > \ln(1) = 0$. Therefore, we just need to show that the numerator

$$\ln\left[\widetilde{\jmath}/\ln(1+\widetilde{\jmath})\right] > 0,$$

or equivalently, that

$$\ln(1+\widetilde{\jmath}) < \widetilde{\jmath}.$$

But this last inequality is equivalent with the inequality

$$1 + \widetilde{\jmath} < e^{\widetilde{\jmath}},$$

which is true for positive $\widetilde{\jmath}$; this may be seen by looking at the Taylor series expansion for e^x around zero, namely

$$e^x = \sum_{k=0}^{\infty} \frac{x^n}{n!} = 1 + x + \sum_{k=2}^{\infty} \frac{x^n}{n!}.$$

In order to show that our critical value for the function $G(f)$ is less than 1, we must establish the inequality

$$\ln\left[\tilde{\jmath}/\ln(1+\tilde{\jmath})\right] < \ln(1+\tilde{\jmath}).$$

Since the natural logarithm is an increasing function, this is equivalent to $\tilde{\jmath}/\ln(1+\tilde{\jmath}) < 1+\tilde{\jmath}$ or to

$$1-\frac{1}{1+\tilde{\jmath}} = \frac{\tilde{\jmath}}{1+\tilde{\jmath}} < \ln(1+\tilde{\jmath}).$$

But, the function $f(x) = 1-\frac{1}{1+x}$ is zero when it is evaluated at $x=0$, as is the function $g(x) = \ln(1+x)$. If $x>0$ $f'(x) = \frac{1}{(1+x)^2} < \frac{1}{1+x} = g'(x)$ for $x>0$. It therefore follows that $f(x) > g(x)$ for $x>0$. In particular,

$$1-\frac{1}{1+\tilde{\jmath}} = f(\tilde{\jmath}) > g(\tilde{\jmath}) = \ln(1+\tilde{\jmath})$$

as desired.

Finally, we are asked to find the maximum when $\tilde{\jmath} = .01$, then when $\tilde{\jmath} = .07$, and finally when $\tilde{\jmath} = .21$ We use the Equation (*), and obtain

$$f = \frac{\ln\left[.01/\ln(1.01)\right]}{\ln(1.01)} \approx 5.0041,\ f = \frac{\ln\left[.07/\ln(1.07)\right]}{\ln(1.07)} \approx .50282,$$

and

$$f = \frac{\ln\left[.21/\ln(1.21)\right]}{\ln(1.21)} \approx .50794,$$

respectively.

(6.8) Yield rate examples

(1) Note that $F = C = \$2{,}000$, $m=1$, $n=N=20$, $r=\alpha=12\%$, and the coupon amount is $Fr = \$240$. The coupons are reinvested at an annual effective rate of $\left(1+\frac{.06}{2}\right)^2 - 1 = 6.09\%$. Therefore, since there is a \$240 coupon received at the end of each year for twenty years, at the time of the last coupon reinvestment, the invested coupons have accumulated value $\$240 s_{\overline{20}|6.09\%} \approx \$8{,}914.434648 \approx \$8{,}914.43$. The investor also receives the redemption amount at the end of the twenty years. Therefore, letting Q denote the price of the bond to yield the investor exactly 8% annually, we have the time 20 equation of value

$$Q(1.08)^{20} = \$8{,}914.43 + \$2{,}000 = \$10{,}914.43.$$

Thus, $Q = \$10{,}914.43(1.08)^{-20} \approx \$2{,}341.671391$. If the price is less than Q, the investor will have a higher yield, but if it is higher than Q, the yield will fall below 8%. Since the investor must pay an integer number of cents, the highest price P that the investor can pay for the bond and obtain an effective yield rate of at least 8% is \$2,341.67.

(3) We have $F = C = \$1{,}000$, $N=8$, $m=2$, $n=Nm=16$, $\alpha = 13\%$, $r = \frac{13\%}{2} = 6.5\%$, $Fr = \$65$, $I = 9\%$, and $j = \frac{9\%}{2} = 4.5\%$. Using the basic price formula,

$$P = \$65 a_{\overline{16}|4.5\%} + \$1{,}000(1.045)^{-16} \approx \$1{,}224.680301 \approx \$1{,}224.68.$$

So, the amount of the premium is $P - C = \$1{,}224.68 - \$1{,}000 = \$224.68$.

The 6% account has a semiannual effective interest rate $q = (1.06)^{\frac{1}{2}} - 1 \approx 2.95630141\%$. Therefore, the equation $\$224.86 = D s_{\overline{16}|q}$ has solution $D \approx 11.18504596$. We round up to $D = \$11.19$, so that the accumulated value is at least the premium amount. Note that the deposits of $D = \$11.19$ accumulate to $\$11.19 s_{\overline{16}|q} \approx \$224.7795144 \approx \$224.78$; this is \$.10 more than the premium.

In order to compute Mitch's yield, as usual we take a bottom-line approach. Counting time in coupon periods (*half years*)

Mitch had an outflow of \$1,224.68 at $t = 0$;

Mitch had net inflows of \$65 – \$11.19 = \$53.81 at times $t = 1, 2, \ldots, 15$;

Mitch had a net inflow of 1,000 + (\$65 – \$11.19) + \$224.78 at $t = 16$.

Thus, the semi-annual yield rate r satisfies the equation of value

$$\$1{,}224.68 = \$53.81 a_{\overline{16}|r} + \$1{,}224.78(1+r)^{-16},$$

and $r \approx 4.394163311\%$. The equivalent annual yield is $(1+r)^2 - 1 \approx 8.981413334\% \approx 8.98141\%$. If you do not note that the savings account will accumulate to ten cents more than the premium, the yield you would obtain is 8.98066%.

(5) We refer to the bond maturing on March 1 of the year $1991 + k$ as bond k, and apply the superscript $[k]$ to its "bond alphabet soup ". We have $F^{[k]} = C^{[k]} = \$1{,}000$, and $m^{[k]} = 1$, for $k \in \{1, 2, 3, 4, 5, 6\}$. If k is odd, so that the bond matures in an even year, then $r^{[k]} = \alpha^{[k]} = 12\%$ and the coupon amount is $F^{[k]} r^{[k]} = \$120$; if k is even, so that the bond matures in an odd year, $r^{[k]} = \alpha^{[k]} = 10\%$ and the coupon amount is $sFr^{[k]} = \alpha^{[k]} = 100$.

On March 1, of 1991 and 1992, Juanita receives coupons for all six bonds, so she receives a total of 3(\$120) + 3(\$100) – \$660. In 1992, she also receives the first \$1,000 redemption, so her total received on March 1, 1992 is \$1,660. Thereafter, each March 1st she receives a \$1,000 redemption payment, and the number of coupons she receives goes down by one. Since the first-redeemed bond has coupon payment $F^{[1]} r^{[1]} = \$120$, the sequence of payments in years 1992 through 1997 is as follows:

(1992) \$1,660,
(1993) \$1,540 = \$1,660 – \$120,
(1994) \$1,440 = \$1,540 – \$100,
(1995) \$1,320 = \$1,440 – \$120,
(1996) \$1,220 = \$1,320 – \$100,
(1997) \$1,100 = \$1,220 – \$120.

Therefore, a March 1, 1990 equation of value is

$$\$6{,}317 = \frac{\$660}{(1+i)} + \frac{\$1{,}660}{(1+i)^2} + \frac{\$1{,}540}{(1+i)^3} + \frac{\$1{,}440}{(1+i)^4} + \frac{\$1{,}320}{(1+i)^5} + \frac{\$1{,}220}{(1+i)^6} + \frac{\$1{,}110}{(1+i)^7}.$$

You may then use "guess and check" or 'Newton's method, but the **Cash Flow worrksheet** of the BA II Plus calculator is ideal ; Enter CF0= –6,317, C01 = 660, C02 = 1,660, C03 = 1,540, C04=1,440, C05 = 1,320, C06 = 1,220, C07 = 1,100, each with frequency 1, then key **IRR** **CPT**. You should find that $i \approx 9.377767799\% \approx 9.37777\%$.

(7) The bond has $F = \$8{,}000$, $C = \$9{,}000$, $N = 15$, $m = 2$, $n = Nm = 30$, $\alpha = 7\%$, $r = \frac{7\%}{2} = 3.5\%$, and coupon amount $Fr = \$280$. Pierre buys the bond at the end of twenty-three months and sells it $7 \times 12 = 84$ months later, that is to say at the end of the 107th month; this is one month before the eighteenth coupon, since $107 = (6 \times 18) - 1$. Thus Irene, who purchases the bond from Pierre, is purchasing a level annuity with $30 - 17 = 13$ payments of \$280, the first of which will be paid in one month, along with the rights to a \$9,000 redemption payment at the time of the last annuity payment. We were told that Irene's price P_{Irene} gave her a $\frac{3.6\%}{2} = 1.9\%$ yield rate per coupon period, so

$$P_{\text{Irene}} = \left[\$280 a_{\overline{13}|1.9\%} + \$9{,}000(1.019)^{-13}\right](1.019)^{\frac{5}{6}} \approx (\$10{,}245.16524)(1.015808447) \approx \$10{,}407.1254.$$

Thus, Irene's and Pierre's common price was \$10,407.13. To find Eric's price P_{Eric}, which was the price of the bond at issue, we use the fact that his yield per coupon period is $\frac{3.6\%}{2} = 1.8\%$. Since Eric held the bond for twenty-three months ($\frac{23}{6}$ *coupon periods*), he received three semiannual \$280 coupons and then \$10,407.13 at the time of sale. Thus,

$$P_{\text{Eric}} = \$280a_{\overline{3}|1.8\%} + [\$10{,}407.13(1.018)^{-\frac{23}{6}}] \approx \$10{,}529.85774 \approx \$10{,}529.86.$$

We might have used \$10,407.12 rather than \$10,407.1 for the price paid by Irene, in which case we would have found Pierre's price to have been \$10,529.85; this is what is reported on page 478 of the text.

(6.9) Callable bonds

(1) (a) Note that $F = C = \$50{,}000$, $m = 1$, $n = N = 10$, $r = \alpha = 8\%$, $P = \$51{,}248$, and the coupon amount is $Fr = \$4{,}000$. The call premiums at times $t = 6$, $t = 7$, $t = 8$, and $t = 9$, are $\$300(10-6) = \$1{,}200$, $\$300(10-7) = \$900$, $\$300(10-8) = \$600$, and $\$300(10-9) = \$300$, respectively; there is no call premium at $t = 10$. Denote the yield rate if the bond is called at time t by y_t; these are yield rates to the buyer who bought the bond at the time it was issued. We have

$$\$51{,}248 = \$4{,}000a_{\overline{6}|y_6} + (\$50{,}000 + \$1{,}200)(1+y_6)^{-6} \quad \text{and} \quad y_6 \approx 7.792348011\%;$$

$$\$51{,}248 = \$4{,}000a_{\overline{7}|y_7} + (\$50{,}000 + \$900)(1+y_7)^{-7} \quad \text{and} \quad y_7 \approx 7.7284444618\%;$$

$$\$51{,}248 = \$4{,}000a_{\overline{8}|y_8} + (\$50{,}000 + \$600)(1+y_8)^{-8} \quad \text{and} \quad y_8 \approx 7.684946616\%;$$

$$\$51{,}248 = \$4{,}000a_{\overline{9}|y_9} + (\$50{,}000 + \$300)(1+y_9)^{-8} \quad \text{and} \quad y_8 \approx 7.654900747\%;$$

$$\$51{,}248 = \$4{,}000a_{\overline{10}|y_{10}} + (\$50{,}000)(1+y_{10})^{-8} \quad \text{and} \quad y_8 \approx 7.634137585\%.$$

Thus, the lowest yield Dominique might receive is about 7.63414%; this is the yield if the bond is held to maturity. The highest yield is earned if the bond is called at the end of six years; it is approximately 7.79235%.

(b) We now assume that any money Dominique receives is immediately deposited into a savings account with a 6% annual effective rate of interest.

• If the bond is called at $t = 6$, then the accumulated value at the end of the ten years is

$$\left[\$4{,}000s_{\overline{6}|6\%} + \$51{,}200\right](1.06)^4 \approx \$99{,}863.53612,$$

and the yield rate is $y_6 = \left(\frac{99{,}863.53612}{51{,}248}\right)^{\frac{1}{10}} - 1 \approx 6.8988423\%.$

• If the bond is called at $t = 7$, then the accumulated value at the end of the ten years is

$$\left[\$4{,}000s_{\overline{7}|6\%} + \$50{,}900\right](1.06)^3 \approx \$100{,}611.4942,$$

and the yield rate is $y_7 = \left(\frac{100{,}611.4942}{51{,}248}\right)^{\frac{1}{10}} - 1 \approx 6.97863883\%.$

- If the bond is called at $t = 8$, then the accumulated value at the end of the ten years is

$$\left[\$4{,}000s_{\overline{8}|6\%} + \$50{,}600\right](1.06)^2 \approx \$101{,}337.3398,$$

and the yield rate is $y_8 = \left(\frac{101{,}337.3398}{51{,}248}\right)^{\frac{1}{10}} - 1 \approx 7.05567449\%$.

- If the bond is called at $t = 9$, then the accumulated value at the end of the ten years is

$$\left[\$4{,}000s_{\overline{9}|6\%} + \$50{,}300\right](1.06) \approx \$102{,}041.1798,$$

and the yield rate is $y_9 = \left(\frac{102{,}041.1798}{51{,}248}\right)^{\frac{1}{10}} - 1 \approx 7.129691673\%$.

- If the bond is called at $t = 10$, then the accumulated value at the end of the ten years is

$$4{,}000s_{\overline{10}|6\%} + \$50{,}000 \approx \$102{,}723.1798$$

and the yield rate is $y_{10} = \left(\frac{102{,}723.1798}{51{,}248}\right)^{\frac{1}{10}} - 1 \approx 7.201078178\%$.

Therefore, assuming Dominique invests all inflows at $i = 6\%$, the lowest yield Dominique might receive is about 6.89884%; it is the yield if the bond is called after six years. The highest yield is earned if the bond is not called in early; it is about 7.20108%.

We note that we did a little more work than was needed. From the accumulated values, it is clear when Dominique gets the highest and lowest yields; it was therefore not necessary that we calculated y_7, y_8, and y_9.

(3) (a) Sofia's eight-year par-value bond has $F = C = \$6{,}000$, $m = 1$, $n = N = 8$, $r = \alpha = 7\%$, and coupon amount $Fr = \$420$. We are given that if the bond is held to maturity, the yield is $y_8 = 6.6\%$. Therefore, to find Sofia's purchase price, we compute

$$P = \$420a_{\overline{8}|6.6\%} + \$6{,}000(1.066)^{-8}.$$

We calculate $P \approx \$6{,}145.559462$, so the price is \$6,145.56.

If the bond is redeemed at $t = 2$ for \$6,300, the yield is y_2 where

$$\$6{,}145.56 = \$420a_{\overline{2}|y_2} + \$6{,}300(1 + y_2)^{-2} \quad \text{and} \quad y_2 \approx 8.042146816\%.$$

If the bond is redeemed at the end of five, six, or seven years, the redemption amount is \$6,200. Since this is less than the purchase amount (*so the buyer had a discount*), the smallest yield (for $t \in \{5, 6, 7\}$) is for $t = 7$; this yield y_7 satisfies

$$\$6{,}145.56 = \$420a_{\overline{7}|y_7} + \$6{,}200(1 + y_7)^{-7} \approx \$6.936762219\%$$

.

Since the yield to maturity y_8 is smaller than y_2 and y_7, the minimal yield is 6.6%; this is realized if the bond is held to maturity.

(b) Note that the intention of the problem was to ask for Sofia's minimal yield rate for the eight-year period. Otherwise, the information that the redemption amount is reinvested (*but the coupons are not reinvested*) is irrelevant. The solution that follows finds minimal yield rate for the eight-year period.

If the bond is redeemed at $t = 2$, then Sofia deposits the redemption amount \$6,300 into the 5.5% account, and it accumulates there for $8 - 2 = 6$ years to $\$6{,}300(1.055)^6$. So, Sofia 's yield y_2 for the eight-year bond period satisfies the time 0 equation of value

$$\$6{,}145.56 = \$420a_{\overline{2}|y_2} + [\$6{,}300(1.055)^6](1 + y_2)^{-8}$$

and $y_2 \approx 6.178073679\% \approx 6.17807\%$; this is most easily found using the **Cash Flow worksheet** with CF0 = −6,145.56, C01=420, F01=2, C02=0, F02=5, C03=$\$6{,}300(1.055)^6$, and F03= 1.

If the bond is redeemed at $t = 5$, then Sofia deposits the \$6,200 redemption amount into the 5.5% account, and it accumulates there for $8 - 5 = 3$ years to $\$6{,}200(1.055)^3$. So, Sofia 's yield y_5 for the eight-year bond period satisfies the time 0 equation of value

$$\$6{,}145.56 = \$420a_{\overline{5}|y_5} + [\$6{,}300(1.055)^3](1 + y_5)^{-8} \quad \textit{and} \quad y_5 \approx 6.496842044\% > 6.17807\%.$$

This is easily obtained with the **Cash Flow worksheet**. (*If you prefer, you do not need to calculate y_5; instead, you may note that with $y_2 \approx 6.178073679\%$,*

$$\$420a_{\overline{5}|y_2} + [\$6{,}300(1.055)^3](1 + y_2)^{-8} > \$6{,}145.56.$$

Hence $y_5 > y_2$.)

Our method for finding the yield rates y_6 and y_7 for the eight-year bond period is similar, but now the \$6,200 redemption payment accumulates to $\$6{,}200(1.055)^2$ and \$6,200(1.055) respectively; we calculate $y_6 \approx 6.653138658\%$ and $y_7 \approx 6.794961825\%$. These are each greater than the computed value for y_2 as is $y_8 = 6.6\%$. (*Once again, the precise values of these yield rates are not required.*)

We conclude that the minimal yield rate for the eight-year bond period is 6.17807%; this is y_2.

(6.10) Floating-rate bonds

(1) Since the face amount is \$2,000, using the given sequence of coupon rates, the successive coupons amounts are .07(\$2,000) = \$140, .064(\$2,000) = \$128, .058(\$2,000) = \$116, .062(\$2,000) = \$124, and .066(\$2,000) = \$132. We are given $P = F = \$2{,}000$, $m = 1$, $n = N = 5$, and $C = \$2{,}125$.

If i is the annual yield rate (*i is equal to the yield rate per coupon period j since $m = 1$*) and $v = \frac{1}{1+i}$, we have the time 0 equation of value

$$\$2{,}000 = \$140v + \$128v^2 + \$116v^3 + \$124v^4 + (\$132 + \$2{,}125)v^5.$$

Consequently, $i \approx 7.492175074\%$. (This may be found quicky using the **Cash Flow worksheet** with CF0 = −2,000, C01=140, C02=128, C03=116, C04=124, C05=132, and F01=F02=F03=F04=F05=1; key **IRR** **CPT**. Alternatively, "Guess and check" or Newton's method may be used.)

Let r denote the level coupon rate which would result in the same yield rate i. Then, the level coupon amount is $Fr = \$2{,}000r$, and we have the time 0 equation of value

$$\$2{,}000 = (\$2{,}000r)a_{\overline{5}|i} + \$2{,}125(1 + i)^{-5}.$$

Thus, using $i \approx 7.492175074\%$,

$$r = \frac{2{,}000 - 2{,}125(1 + i)^{-5}}{2{,}000a_{\overline{5}|i}} \approx \frac{519{,}2740744}{8{,}093.452081} \approx 6.415977623\% \approx 6.41598\%.$$

(6.11) The BAII Plus calculator Bond worksheet

(1) Note that $F = \$1{,}500$, $m = 1$, $r = \alpha = 5\%$, $C = \$1{,}650$, and the yield to the new buyer is $\tilde{j} = 8.2\%$.

To calculate the price paid at the February 23rd settlement and the accrued interest included in the price, using the **Bond Worksheet**, we need entries "SDT = 2 - 23 - 1995", "CPN = 5", "RDT = 8 - 01 - 1995", "RV = 110" (since $\frac{C}{F} \times 100 = \frac{165}{150} \times 100 = 110$) , and "YLD = 8.2" Furthermore, on the worksheet you must select a "360" basis (not an "ACT" basis) and the coupons set to "1/Y" (rather than "2/Y"). Then, you may scroll to the PRI register and key **CPT** to obtain "PRI = 108.199496". Store this displayed value and press $\boxed{\downarrow}$ to obtain "AI = 2.805555556". If you now add this to your stored "PRI" value, you obtain 111.0050515, and this represents the price (in dollars) per \$100 of face value. Therefore, the price of the \$1,500 bond was $\left(\frac{\$1,500}{100}\right)(111.0050515) \approx \$1,665.075773 \approx \$1,665.08$. The amount of accrued interest was $\left(\frac{\$1,500}{100}\right)(2.805555556) \approx \$42.08333333 \approx \$42.08$.

Without the **Bond worksheet**, begin by observing that on a "30/360" basis, there are

$$(23-1)+(8-2)(30)+(1995-1994)(360)=202$$

days between August 1, 1994 (*the coupon date prior to the resale*) and February 23, 1995, so $f=\frac{202}{360}$ is the fraction of the year that the old buyer held the bond. At the time of resale, the bond has less than one coupon period remaining, and thus Fact (6.11.2) applies; the clean price used is

$$\begin{aligned}\frac{C+Cg}{1+(1-f)\jmath}-fCg &= \frac{C+Fr}{1+(1-f)\jmath}-fFr\\ &= \frac{\$1,650+\$75r}{1+\left[1-\left(\frac{202}{360}\right)\right].082}-\left(\frac{202}{360}\right)(\$75)\\ &\approx \$1,665.075773-\$42.083333 \approx \$1,622.99244.\end{aligned}$$

Here, we have subtracted off \$42.08, and this is the accrued interest. The invoice (dirty price) is \$1,665.08.

(3) We have $F=C=\$3,000$, $m=2$, $\alpha=6\%$, and $r=3\%$. The coupon amount is $Fr=\$90$. Since the bond is a government bond, we will use an "actual/actual" basis. Open the **Bond worksheet** and then set "SDT = 8 - 11 - 02", "CPN = 6", "RDT = 12 - 12 - 02", "RV=100". Next select "ACT" and "2/Y". Fix "YLD=4.5". If you then scroll to the PRI register and key $\boxed{\textbf{CPT}}$, you should obtain "PRI = 100.4942605". Then, pressing $\boxed{\downarrow}$ gives you "AI = .93442623". The sum of the entries in the PRI and AI registers is about 101.4286867 , so, the invoice price to allow the buyer a yield of 4.5% is

$$\left(\frac{\$3,000}{100}\right)(101.4286867) \approx \$3,042.860602 \approx \$3,042.86.$$

The accrued interest in the price is

$$\left(\frac{\$3,000}{100}\right)(.934442623) \approx \$28.03278689 \approx \$28.03.$$

We now wish to check this value for the invoice price without using the worksheet. We first calculate that the number of days between August 11, 2002 and December 15, 2002 is

$$(31-11)+30+31+30+15=126.$$

There are

$$(30-15)+31+31+30+31+30+15=183$$

days between June 15, 2002 and December 15, 2002, so $f=\frac{183-126}{183}=\frac{57}{183}$. We are interested in an invoice price so that the buyer will earn 4.5% semi-annually, so $\jmath=\frac{4.5\%}{2}=2.25\%$. There is less than one coupon period until redemption, so we use Fact (6.11.2):

$$\begin{aligned}\frac{C+Cg}{1+(1-f)\jmath}-fCg &= \frac{C+Fr}{1+(1-f)\jmath}-fFr\\ &= \frac{\$3,090}{1+\left[1-\left(\frac{57}{183}\right)\right)\left(\frac{126}{183}\right]}-\left(\frac{157}{183}\right)(\$90)\\ &\approx \$3,042.860602 \approx \$3,042.86.\end{aligned}$$

The accrued interest in the price is $Cgf = Frf = \$90\left(\frac{157}{183}\right) \approx \$28.03278689 \approx \$28.03$. So, there is complete agreement in the numerical results obtained by the two methods.

Chapter 6 review problems

(1) Tabitha's bond has $F = C = \$1{,}000$, $m = 1$, $g = r = \alpha = 6\%$, $j = I = 5\%$, and coupon amount $Fr = \$60$. Note that the bond is given to be an n-year bond; since the bond has annual coupons, as usual n is the number of coupons the bond pays. We are given $I_1 = \$52.89$. But, by Equation (6.5.6),

$$I_1 = Cg - C(g-j)v_j{}^{n-1+1} = \$60 - \$1{,}000(.06 - .05)(1.05)^{-n}.$$

Therefore, $\$60 - \$52.89 = \$10(1.05)^{-n}$, $(1.05)^n = \frac{10}{7.11}$, and

$$n = \ln\left(\frac{10}{7.11}\right) \Big/ \ln(1.05) \approx 6.990811672.$$

Since n must be an integer, as it is the number of coupons in the bond, $n = 7$. Then, Equation (6.3.3) (*which is an immediate consequence of the premium-discount formula*) gives the amount of premium to be

$$P - C = \$1{,}000(.06 - .05)a_{\overline{7}|5\%} \approx \$57.86373 \approx 57.86.$$

(3) We have $F = \$4{,}000$, $m = 4$, $\alpha = 10\%$, $r = \frac{10\%}{4} = 2.5\%$, $C = \$3{,}800$, and $K = \$1{,}950.82$. Moreover, $i^{(2)} = 8\%$ and $j = i^{(4)}/4 = \left(1 + \frac{.08}{2}\right)^{\frac{1}{2}} - 1 \approx .019803903$. The coupon amount is $Fr = (\$4{,}000)(.0225) = \$100$; it is also equal to Cg so $g = \frac{\$100}{\$3{,}800} \approx .026315789$. Using Makeham's formula [Equation (6.4.4)],

$$P = \frac{g}{j}(C-K) + K \approx \frac{.026315789}{.019803903}(\$3{,}800 - \$1{,}950.82) + \$1{,}950.82 \approx \$4{,}408.044329 \approx \$4{,}408.04.$$

To calculate the number of coupons n, we note that

$$\$1{,}950.82 = K = Cv_j{}^{-n} = \$3{,}800\left[\left(1 + \frac{.08}{2}\right)^{\frac{1}{2}}\right]^{-n} = \$3{,}800(1.04)^{-\frac{n}{2}}.$$

Therefore, $(1.04)^{\frac{n}{2}} = \frac{3{,}800}{1{,}950.82}$, and $n = \frac{2\ln(3{,}800/1{,}950.82)}{\ln(1.04)} \approx 33.99995646$. The number of coupons is an integer, and therefore it is 34.

Alternatively, you might use the premium-discount formula $P = C(g-j)a_{\overline{n}|j} + C$; from the equation

$$\$4{,}408.04 \approx \$3{,}800(.026315789 - .019803903)a_{\overline{n}|1.9803903\%} + \$3{,}800,$$

we find

$$a_{\overline{n}|1.9803903\%} \approx \frac{4{,}408.04 - 3{,}800}{3{,}800(.026315789 - .019803903)} \approx 24.57206833,$$

and $n \approx 33.99961229$. Once again, the number of coupons is an integer, and therefore it is 34.

Yet another possibility is to use the basic price formula with the **TVM worksheet** to obtain 34 as the number of coupons.

(5) For each bond, $F = C = \$2{,}000$, $m = 1$, $r = \alpha = 7.3\%$, and coupon amount is $Fr = \$146$. When Mr. Cole holds all eight bond, the total coupon payment is $8 \times \$146 = \$1{,}168$. When a bond is redeemed, there is a \$2,000 redemption payment, but for each bond that has been redeemed in a prior year, there is a reduction of \$146 in the total coupon payments received. Thus, Mr. Cole receives \$1,168 on October 30, 1979, $\$1{,}168 + \$2{,}000 = \$3{,}168$ on October 30, 1980, and after that Mr. Cole's

receipts decrease by a total of \$146 each year, terminating with a payment of \$2,146 on October 30, 1978.

Mr. Cole's yield for the nine-year period is 8%, so we have

$$\begin{aligned} P &= \$1{,}168(1.08)^{-1} + (1.08)^{-1}\left[(I_{\$3{,}168,-\$146}a)_{\overline{8}|8\%}\right] \\ &= \$1{,}168(1.08)^{-1} + (1.08)^{-1}\left[\$3{,}168a_{\overline{8}|8\%} - \frac{\$146}{.08}\left(a_{\overline{8}|8\%} - 8(1.08)^{-8}\right)\right] \\ &\approx \$1{,}081.481481 + (1.08)^{-1}(\$18{,}205.35217 - \$2{,}599.6903590 \\ &\approx \$15{,}531.16835. \end{aligned}$$

Thus, the price of the portfolio is \$15,531.17.

(7) We have $F = C = \$5{,}000$, $P = \$5{,}250$, $m = 2$, $N = 8$, $n = mN = 16$, $\alpha = 6.75\%$, $r = \frac{6.75\%}{2} = 3.375\%$, and coupon amount $Fr = \$168.75$. Whether or not the put option is exercised at the end of five years does not alter the cashflows stemming from coupon payments during the five years. Rather, the exercise or lack of exercise just determines whether the investor receives a single redemption payment of \$5,000 at time 5, which grows to an amount $\$5{,}000(1+I)^3$ at time 8, or six \$168.75 coupon payments at times $5\frac{1}{2}$, 6, $6\frac{1}{2}$, 7, $7\frac{1}{2}$, and 8, along with the \$5,000 redemption at time 8. In the latter case, these accumulate to $\$168.75s_{\overline{6}|(1+I)^{\frac{1}{2}}-1} + \$5{,}000$ at time 8. Thus, the investor should should prefer to exercise the option if

$$\$5{,}000(1+I)^3 > \$168.75s_{\overline{6}|(1+I)^{\frac{1}{2}}-1} + \$5{,}000.$$

Note that this inequality is equivalent with

$$\$5{,}000(1+I)^3 > \$168.75\left[\frac{(1+I)^3 - 1}{(1+I)^{\frac{1}{2}} - 1}\right] + \$5{,}000,$$

which in turn is equivalent to

$$\$5{,}000\left[(1+I)^3 - 1\right] > \$168.75\left[\frac{(1+I)^3 - 1}{(1+I)^{\frac{1}{2}} - 1}\right],$$

and to

$$(1+I)^{\frac{1}{2}} - 1 > \frac{168.75}{5{,}000} = .03375.$$

Therefore, the purchaser should exercise the put option if $I > (1.03375)^2 - 1 \approx 6.86390625\%$.

A shorter way to obtain this same result is to note that if he holds onto the bond until time 8, each year he receives coupons paying a nominal return of 6.75% convertible semiannually on the \$5,000 investment amount, hence an effective return of $(1.03375)^2 - 1 \approx 6.86390625\%$. If he exercises the put option, his return in years 6 through 8 is I. So, he should exercise the put option if $I > 6.86390625\% \approx 6.86391\%$.

CHAPTER 7

Stocks and financial markets

(7.1) Common and preferred stock

(1) Since Bridget purchased \$200 shares and the stock pays \$.28 per share each quarter, she will receive $200 \times \$.28 = \$56.00$ at the end of each quarter. The price is based on an effective quarterly interest rate of $J = (1.062)^{\frac{1}{4}} - 1$ and the premise that the dividends continue forever. Therefore, the price of the \$200 shares is

$$P = \frac{\$56}{J} = \frac{\$56}{(1.062)^{\frac{1}{4}} - 1} \approx \$3{,}695.8506 \approx \$3{,}695.85.$$

Thus, Bridget paid \$3,695.85 for 200 shares and her per share price is $\frac{\$3{,}695.85}{200} \approx \$18.47925 \approx \$18.48$.

Note, that we could also have found the per share price by calculating $\frac{\$.28}{(1.062)^{\frac{1}{4}}-1} \approx 18.49253 \approx$ \$18.48, since a single unit of stock gives rise to a perpetuity paying \$.28 at the end of each quarter.

(7.2) Brokerage accounts

(1) The initial margin requirement is 65%, so Dr. Rogowski deposits $.65 \times \$13{,}000 = \$8{,}450$ and borrows $\$13{,}000 - \$4{,}550$ from her broker. So, the market value X of the stocks is initially \$4,550 more than the \$8,450 equity in the account. According to the maintenance margin requirement, this equity must be at least 45% of the market value; that is $X - \$4{,}550 \geq .45X$ is required. Equivalently, the maintenance margin requirement calls for $X > \frac{\$4{,}550}{.55} \approx \$8{,}272.7273$. Therefore, there is a margin call if the price goes down to \$8,272.72. Since, $\frac{8{,}272.72.72}{13{,}000} \approx .636363077 = 1 - .363636923$, there is a margin call if the price of the stock goes down by about 36.36370% (*or more*).

(3) (a) Beverly purchased \$20,000 of stock using \$10,000 of her own money and \$10,000 of borrowed money. When the price fell to \$15,000, her debit balance remained at \$10,000, but she then had only \$5,000 of equity. The maintenance margin requires that she have equity of at least $.35 \times \$15{,}000 = \$5{,}250$, so Beverly must deposit \$250.

(b) Suppose that Beverly deposits securities having a market value S in her account. Then, her equity will be $\$5{,}000 + S$ while the value of the securities in her account will now be $\$15{,}000 + S$. Therefore, in order for the 35% margin requirement to be satisfied, we need

$$\$5{,}000 + S \geq .35 \times (\$15{,}000 + S).$$

This is equivalent with the inequality $.65S \geq \$250$, so we need $S \geq \frac{\$250}{.65} \approx \$384.6153846$. We therefore need a deposit of at least \$384.62 in securities , assuming this is Beverly's only response.

(c) Suppose that L of stocks are sold and the money from the sale is added to the margin account. Then, there remains \$5,000 of equity, and this must be at least 50% of the value of the remaining securities; that is, we need $\$5{,}000 \geq .5 \times (\$15{,}000 - L)$. Equivalently, $L \geq \$5{,}000$, so at least \$5,000 of the stock must be sold.

(7.3) Going long: buying stock with borrowed money

(1) Note that $\frac{\$3{,}000}{35.45} \approx \$84.626$. So, Eleanor can buy $84 shares of stock using $84 \times \$35.45 = \$2{,}977.80$ of cash. On the other hand, If she uses a margin account with a 60% margin account, the funds available for her stock purchase are $3{,}000/.06 = \$5{,}000$, and since $\frac{\$5{,}000}{35.45} \approx 141.0437236$, 141 shares may be bought.

Suppose that three months later, the per share price is \$38.45 and she sells all the shares she bought using a cash account. if she was using a cash account, she sells 84 shares for $\frac{\$3{,}000}{38.45} = \$3{,}229.80$ and her yield j for the three month period satisfies $\$2{,}977.80(1+j) = \$3{,}229.80$. Thus, $j \approx 8.46262\%$. [*Her annual yield rate* $i = (1+j)^4 - 1 \approx 38.3950059\%$, *but the problem did not ask for this.*]

Again, suppose that three months later the per share price is \$38.45 and she liquidates her stocks, but now suppose that this consists of 141 shares, purchased using a margin account to which she had deposited \$3,000. Then, since the price of the stock went up by \$3 per share and \$30 of interest had to be paid by Eleanor, she has a profit of $(141 \times \$3) - \$30 = \$423 - \$30 = \$393$. Therefore, her quarterly yield rate is equal to $\frac{393}{3{,}000} = 13.1\%$.

Note that in order to purchase 141 shares, she would only have been required to invest $(.6)(141)(\$35.45) = \$2,999.07$. If this was what she elected to do, her profit would still have been \$393, and her quarterly yield would have been $\frac{393}{2{,}999.07} = 13.1040623\% \approx 13.1041\%$.

(7.4) Selling short: selling borrowed stocks

(1) With notation as introduced in section (7.4), we have $P_0 = P$, $P_1 = .9P$, $J = 5\% = .05$, $r = 50\% = .5$, and $i_{\text{shortsale}} = 14\% = .14$. Therefore, Equation (7.4.2) tells us that

$$.14 = \frac{(P - .9P) + (.05)(.5P) - D}{.5P} = \frac{.125P - D}{.5P}.$$

It follows that $.07P = .125P - D$, so $D = .055P$. The dividends are 5.5% of P, or 11% of his original investment of $.5P$.

(3) With notation as introduced in Section (7.4), we have $P_0 = \$2{,}400$, $P_1 = \$2{,}012$, $J = 6\% = .06$, $r = q\% = \frac{q}{100}$, $D = \$0$, and $i_{\text{shortsale}} = 34\% = .34$. Using Equation (7.4.2) , we find

$$.34 = \frac{(\$2{,}400 - \$2{,}012) + (.06)\left(\frac{q}{100}\right)(\$2{,}400)}{\left(\frac{q}{100}\right)(\$2{,}400)};$$

so $8.16q = 388 + 1.44q$, and $q = \frac{388}{6.72} \approx 57.738095420$. So, the margin requirement is about 57.73810%.

(5) The purchase price of the stock is \$3,000, so Warren's initial margin contribution is for $.5 \times \$3{,}000 = \$1{,}500$. Dividends of \$20 at the end of six months and of \$15 at the end of eighteen months are deducted from his 4% margin account, resulting in a balance of

$$\$1{,}500(1.04)^{\frac{20}{12}} - \$20(1.04)^{\frac{20-6}{12}} - \$15(1.04)^{\frac{20-18}{12}} \approx \$1{,}565.292702 \approx \$1{,}565.29.$$

When Warren liquidates, he receives

$$\$1{,}565.29 + (\$3{,}000 - \$2{,}650) = \$1{,}915.29.$$

Thus, Warren's annual yield satisfies $\$1{,}500(1+i)^{\frac{20}{12}} = \$1{,}915.29$, and $i \approx 15.79407474\% \approx 15.79407\%$.

Chapter 7 review problems

(1) Since the price of the stock at the time the short sale was initiated was \$2,082, and there was a 60% initial margin requirement, Fernando's margin deposit is for .6(\$2,082) = \$1,249.20. The price of the stock goes up by 2% between the time Fernando sells the borrowed stock and the time he buys it back; this results in a loss of .02(\$2,082) = \$41.64. The accumulated value of the dividends at the end of the year (*the interval between the sale and the repurchase*), figured at the 3.2% annual effective interest rate, is

$$\$84(1.032)^{\frac{1}{2}} \approx \$85.33341667.$$

Finally, the interest on \$1,249.20 for one year, again at an annual effective interest rate of 3.2%, is .032(\$1,249.20) ≈ \$39.9744. Therefore, there is a total <u>loss</u> of approximately

$$\$41.64 + \$85.33341667 - \$39.9744 = \$86.99901667.$$

But the short sale required a \$1,249.20 investment , so using this value for the loss, we calculate that the yield rate for the year is

$$-\$86.99901667/\$1{,}249.20 \approx -.069643785 \approx -6.96438\%.$$

If you use \$87 (*instead of* \$86.99901667), the calculated yield rate is

$$-\$87/\$1{,}249.20 \approx -.06964457253 \approx -6.96446\%.$$

(3) (a) The initial price of the purchased stock is \$16,850. It is purchased using a newly created margin account with a 50% initial margin requirement, so Yuri must deposit $.5 \times \$16{,}850 = \$8{,}425$. Thus, \$8,425 is Yuri's initial equity. When the stock price rose to \$18,200, this was an increase of $\$18{,}200 - \$16{,}850 = \$1,350$, and hence the equity goes up to $\$8{,}425 + \$1{,}350 = \$9{,}775$; the equity is $(\$18{,}200 + \$8{,}425) - \$16{,}850 = \$9{,}775$. This balance is required to be at least 50% of the price of the securities, so at that point Yuri was allowed to withdraw $\$9{,}775 - .5(\$18{,}200) = \$675$. The problem indicates he withdrew the full \$675, leaving a cash balance of $\$8{,}425 - \$675 = \$7{,}750$ in the margin account (along with the stock). The stock price then had a sudden drop to \$13,600, and at this point Yuri received a margin call. At the time of the margin call, he needs equity of at least $.4(\$13{,}600) = \$5{,}440$, and his equity was $(\$13{,}600 + \$7{,}750) \quad (\$16{,}850) - \$4,500 - \$940$. The margin call is therefore for $5{,}440 - \$4,500 = \$940$.

(b) If Yuri adds marginable securities, their value must be S where

$$\$4{,}500 + S = .4(\$13{,}600 + S) = \$5{,}440 + S.$$

Equivalently, we need $S = \$940/.6 \approx \$1{,}566.666667$. Therefore, he adds \$1,566.67 of securities.

(c) Suppose Yuri sold stocks from the margin account valued at Y. His equity would be unchanged by the sale; it would continued to be \$4,500. After the sale of stocks valued at Y, the value of the marginable securities in the account was $\$13{,}600 - Y$, and it was required that Y be as small as possible so that $\$4{,}500 \geq .5(\$13{,}600 - Y) = \$6{,}800 - (.5)(\$Y)$. Thus, we need $Y = \$2{,}300/.5 = \$4{,}600$.

Note: The answer in the back is incorrect. It was based on needing $\$4{,}500 = .4(\$13{,}600 - \$Y) = \$5{,}440 - (.4)(\$Y)$; *then* $\$Y = 940/.4 = \$2{,}350$. *But, we need to meet the initial margin requirement if we are selling off securities, not the maintenance maintenance margin requirement.*

CHAPTER 8

Arbitrage, the term structure of interest rates, and derivatives

(8.2) Arbitrage

(1) Mr. Ralbracht may borrow \$20,000; two years later, he will then need to repay $\$20{,}000(1.025)^2 =$ \$21,012.50. He may use the \$20,000 of borrowed funds to purchase two-year Treasury notes. These have a six-month interest rate of $(1.0372)^{\frac{1}{2}} - 1$, and therefore, assuming he uses exactly \$20,000 to purchase the notes, Mr. Ralbrecht would receive interest payments of $\$20{,}000\left[(1.0372)^{\frac{1}{2}} - 1\right] \approx$ \$368.60 after six months, twelve months, eighteen months, and twenty-four months. At the time of the last interest payment, he will also receive a \$20,000 redemption payment, so his total inflow at the end of two years from the Treasury note would be \$20,368.60. In order that Mr. Ralbracht have an arbitrage opportunity, he must not have a net outflow at any time. Therefore, he must invest a sufficient amount of the prior interest payments, so that at time two years he may withdraw at least $\$21{,}012.50 - \$20{,}368.60 = \$643.90$; if he makes level deposits at times six months, twelve months, and eighteen months, deposits of \$214.64 will clearly suffice since $3\times 214.64 = \$643.92$, but assuming the account pays interest, somewhat smaller deposits will be also suffice. Of course, Mr. Ralbracht may choose to save up the \$643.90 by non-level deposits if he prefers. However, since an arbitrage opportunity cannot have any chance of a net outflow occurring, he must do so in a manner that does not count on any particular interest rate being available in the future.

(3) Let P_k denote the amount of the k-th option that Leonard buys; P_k is negative if he sells the option. Since we don't want Leonard to risk any money, we need $P_1 + P_2 + P_3 \leq 0$. Furthermore, we must be sure that Leonard will not experience a loss if the market goes up or if it goes down. Therefore, since the value of Leonard's holdings is $1.1P_1 + 1.2P_2 + 1.16P_3$ if the market the market is up and $.9P_1 + .85P_2 + .8P_3$ if the market is down, we need to select the set of prices so that we simultaneously have

$$\begin{cases} P_1 + P_2 + P_3 \leq 0 \\ 1.1P_1 + 1.2P_2 + 1.16P_3 \geq 0 \\ .9P_1 + .85P_2 + .8P_3 \geq 0. \end{cases}$$

While you might discover a solution to the above system of simultaneous inequalities by trial and error, we wish to present a systematic method by which you may discover solutions. We note that we have a system of three simultaneous inequalities, and that the corresponding equalities are each linear and represent planes through the origin in $P_1P_2P_3$ coordinates. (*You are probably more used to xyz coordinates, but we can call our variables whatever we like.*) Moreover, the matrix

$$\begin{pmatrix} 1 & 1 & 1 \\ 1.1 & 1.2 & 1.16 \\ .9 & .85 & .8 \end{pmatrix}$$

is invertible since its determinant is nonzero; it is

$$1\left[(1.2)(.8)-(1.16)(.85)\right]-1\left[(1.1)(.8)-(1.16)(.9)\right]+\left[(1.1)(.85)-(1.2)(.9)\right]=-.007.$$

Thus the three planes are distinct and the inequalities determine a region of plane; there are infinitely many solutions.

Let us go ahead and see whether we can find a solution with $P_1+P_2+P_3=0$; this amounts to our asking that there be no initial net inflow as well as no initial net outflow. Then, we seek P_2 and P_3 with

$$\begin{cases} 0\le 1.1(-P_2-P_3)+1.2P_2+1.16P_3=.1P_2+.06P_3 \\ 0\le .9(-P_2-P_3)+.85P_2+.8P_3=-.05-.1P_3. \end{cases}$$

or equivalently

$$\begin{cases} 0\le .1P_2+.06P_3 \\ 0\le -.1P_2-.2P_3. \end{cases}$$

Adding together the two inequalities of this last system, we see that $0\le -.14P_3$, so look for negative P_3. For example, take $P_3=-\$1{,}000$. With this choice of P_3, our last system of inequalities gives

$$\begin{cases} 0\le .1P_2-.\$60 \\ 0\le -.1P_2+\$200. \end{cases}$$

The first equation is equivalent with $\$600\le P_2$, and the second equation is equivalent to $P_2\le\$2{,}000$. So, we can may choose P_2 from the interval $[\$600,\$2{,}000]$. Let's choose $P_2=\$800$. Finally, we need to select P_1. If we go ahead and insist that $P_1+P_2+P_3=0$, we have $P_1=\$200$.

Now let's verify that the simultaneous choices $P_1=\$200$, $P_2=\$800$, and $P_3=-\$1{,}000$ really do give us arbitrage. If the market goes up, Leonard's position is worth $1.1(\$200)+1.2(\$800)+1.16(-\$1{,}000)=\$20$. If the market goes down, his position is worth $.9(200)+.85(\$800)+.8(-\$1{,}000)=\$60$.

Among Leonard's arbitrage opportunities is that he buys \$200 of the first option and \$800 of the second option, while simultaneously selling \$1,000 of the last option. This will result in no loss or gain now, but he is certain of a gain, assuming the market will either go up or down. We want to stress that this is just *one* of the *infinitely many solutions.*

(8.3) The term structure of interest rates

(1) We first compute the spot rates. Since the bonds were zero-coupon bonds, we have

$$\$9{,}765=\frac{\$10{,}000}{1+r_1}\quad\text{and}\quad r_1=\frac{10,000}{9,765}-1\approx .02406554\approx 2.40655\%,$$

$$\$9{,}428=\frac{\$10{,}000}{(1+r_2)^2}\quad\text{and}\quad r_2=\left(\frac{10{,}000}{9,428}\right)^{\frac{1}{2}}-1\approx .02988851\approx 2.98885\%,$$

and

$$\$8{,}986.82=\frac{\$10{,}000}{(1+r_3)^3}\quad\text{and}\quad r_3=\left(\frac{10,000}{8,986.82}\right)^{\frac{1}{3}}-1\approx .036250259\approx 3.62503\%.$$

These findings may also be reported as forward rates:

$$f_{[0,1]} \approx 2.40655\%;\ f_{[0,2]} \approx 2.98885\%;\ f_{[0,3]} \approx 3.62503\%.$$

The remaining forward rates are found as follows, using Equation (8.3.8):

$$f_{[1,2]} = \frac{(1+r_2)^2}{1+r_1} - 1 \approx \frac{(1.02988851)^2}{1.02406554} - 1 \approx 3.57446\%;$$

$$f_{[1,3]} = \left(\frac{(1+r_3)^3}{1+r_1}\right)^{\frac{1}{3-1}} - 1 \approx \left(\frac{(1.036250259)^3}{1.02406554}\right)^{\frac{1}{2}} - 1 \approx 4.23969\%;$$

$$f_{[2,3]} = \frac{(1+r_3)^3}{(1+r_2)^2} - 1 \approx \frac{(1.036250259)^3}{(1.02988851)^2} - 1 \approx 4.90919\%.$$

(3) A two-year \$1,000 6% par-value bond with semiannual coupons has coupon amount

$$Fr = (\$1{,}000)\left(\frac{.06}{2}\right) = \$30$$

and redemption payment \$1,000. The coupons are paid at the end of each six-month period, and the redemption payment is paid at the time of the last coupon. Therefore, using discount factors as determined by the zero-coupon (*pure discount*) bonds, the arbitrage-free price of a two-year \$1,000 6% par-value bond with semiannual coupons is

$$\frac{\$30}{(1.019)^{\frac{1}{2}}} + \frac{\$30}{1.023} + \frac{\$30}{(1.0265)^{\frac{3}{2}}} + \frac{\$1,030}{(1.035)^2} \approx \$1{,}057.822203 \approx \$1{,}057.82.$$

(5) We first use bootstrapping to determine $(1+r_3)^3$ and $(1+r_5)^5$; this is useful because then we can use Equation (8.3.8) to determine $f_{[3,5]}$. We can use bonds with annual coupons having a face value F of our choosing, because the choice of face value will alter all cashflows by the same factor and we are not rounding. We use \$100 par-value 4% bonds with annual coupons, and denote the price of the k year bond to give us the specified yield y_k by P_k, $k \in \{1,2,3,4,5\}$. The coupon amount of these bonds is \$4, and the redemption amount is \$100. Therefore, since $r_1 = y_1 = 1.435\%$,

$$\frac{4}{1.01435} + \frac{\$104}{(1+r_2)^2} = \frac{\$4}{1+r_1} + \frac{\$104}{(1+r_2)^2} = P_2 = \frac{\$4}{1.02842} + \frac{\$104}{(1.02842)^2}, \quad \text{and} \quad (1+r_2)^2 \approx 1.058228304.$$

Using the just-computed value of $(1+r_2)^2$, we have

$$\frac{\$4}{1.01435} + \frac{\$4}{(1+r_2)^2} + \frac{\$104}{(1+r_3)^3} = P_3 = \frac{\$4}{1.03624} + \frac{\$4}{(1.03624)^2} + \frac{\$104}{(1.03624)^3}, \quad \text{and} \quad (1+r_3)^3 \approx 1.114354042.$$

Then, using our values for $(1+r_2)^2$ and for $(1+r_3)^3$, we note

$$\frac{\$4}{1.01435} + \frac{\$4}{(1+r_2)^2} + \frac{\$4}{(1+r_3)^3} + \frac{\$104}{(1+r_4)^4} = P_4 = \frac{\$4}{1.03943} + \frac{\$4}{(1.03943)^2} + \frac{\$4}{(1.03943)^3} + \frac{\$104}{(1.03943)^4}$$

and

$$(1+r_4)^4 \approx 1.169928201.$$

Continuing in this manner, but now using the already-computed values for $(1+r_2)^2$, $(1+r_3)^3$, and $(1+r_4)^4$, we obtain

$$\begin{aligned}
&\frac{\$4}{1.01435}+\frac{\$4}{1.058228304}+\frac{\$4}{1.114354042}+\frac{\$4}{1.169928201}+\frac{\$104}{(1+r_5)^5}\\
&=\left(\sum_{k=1}^{4}\frac{\$4}{(1+r_k)^k}\right)+\frac{\$104}{(1+r_5)^5}\\
&=P_5\\
&=\frac{\$4}{1.04683}+\frac{\$4}{(1.04683)^2}+\frac{\$4}{(1.04683)^3}+\frac{\$4}{(1.04683)^4}+\frac{\$104}{(1.04683)^5},
\end{aligned}$$

and therefore

$$(1+r_5)^5 \approx 1.263899501.$$

Equation (8.3.3) with $t=3$ and $s=5$ now gives us

$$(1+f_{[3,5]})^{5-2}=\frac{(1+r_5)^5}{(1+r_3)^3}\approx\frac{1.263899501}{1.114354042},$$

so

$$f_{[3,5]}\approx\left(\frac{1.263899501}{1.114354042}\right)^{\frac{1}{2}}-1\approx .064987904\approx 6.4987\%.$$

(7) In Example (8.3.3), we calculated that $1+r_1=\frac{1}{.975}=\frac{30}{29.25}$. Therefore, you could sell a one-year zero coupon bond with \$30 redemption for \$29.25. Note that the two-year coupon bond of Example (8.3.3) had price \$984 and that $\$984-\$29.25=\$954.75$. If the two-year spot rate were $r_2=3.5\%$, you could sell a two-year zero coupon bond with redemption amount $\$954.75(1.035)^2\approx\$1{,}022.75$ for \$954.75.

Now, suppose that you sell a one-year zero-coupon bond with \$30 redemption for \$29.25 and a two-year zero coupon bond with redemption amount \$1,022.75; you will obtain $\$29.25+\$954.75=\$984$. Use this \$984 to purchase the two-year coupon bond of Exercise (8.3.3) which had \$30 annual coupons and a \$1,000 redemption amount. The coupon you receive at the end of one year will allow you to exactly meet your obligation to pay the redemption payment for the one-year zero-coupon bond you sold. Moreover, your inflow at the end of the second year will be $\$30+\$1{,}000=\$1{,}030$; after you pay out the \$1,022.75 redemption payment for the two-year zero-coupon bond you sold, you will have $\$1{,}030-\$1{,}022.75=\$7.25$ remaining. You therefore have a way to make money (*at* $t=2$) without having any possibility of a net outflow at any time; this is an arbitrage opportunity.

(8.4) Forward contracts

(1) Let S_0 denote the cost of 10,000 ounces of high grade copper at the time the contract was entered, and note that the cost of holding 10,000 ounces of copper for six months is \$40. So, the buyer Nadia of the forward contract might have paid $S_0+\$40$ so as to ensure she would hold the copper in six months, rather than having entered the contract and having agreed to buy the copper in six months for \$16,800. Since the risk-free rate of interest at that time was 3% convertible quarterly, the risk-free present value of \$16,800 to be paid in six months is $\$16{,}800/(1+\frac{.03}{4})^2$. So, by the "law of one price",

$$\$16{,}800\Big/\left(1+\frac{.03}{4}\right)^2=S_0+\$40,$$

and $S_0\approx\$16{,}510.80691\approx\$16{,}510.81$. At the time the contract was entered, the per-ounce price of the copper was $\$16{,}510.81/10{,}000\approx\$1.65$.

(8.5) Commodity futures held until delivery

(1) Boyd has a positive profit if, at the settlement date, he buys the underlier for an amount that is below the market price, and he has a loss (*negative profit*) if he must buy the underlier above its then current price. The amount B of his profit is the amount (*possibly negative*) by which the spot price S at settlement time exceeds the settlement price K; $B = S - K$. Linda's situation is the opposite of Boyd's. She profits if she sells the underlier above market value and loses if she is obliged to sell below market value; her profit (*possibly negative*) is $L = K - S$. Thus, $B + L = (S - K) + (K - S) = (S - S) + (K - K) = \$0$.

(3) Katrina's initial margin balance is \$800. Since she took a long position, she has purchased the right to buy the underlier on the deliery date for \$6,150. One day later, her balance is to be adjusted assuming that the delivery price is now \$75 higher (since \$6,225 − \$6,150 = \$75) , so she has a better deal by \$75. This \$75 is added to her margin balance, giving her a new margin balance of \$800 + \$75 = \$875.

(8.6) Offsetting positions and liquidity of futures contracts

(1) The price fell from \$19,120 to \$18,310. Since $1 - \$18,310/\$19,120 \approx .042364017$, the price fell by about 4.23640%. Trevor made a deposit of \$2,700, and this grew to

$$\$2{,}700\left(1 + \frac{.027}{12}\right)^2 \approx \$2{,}712.163669 \approx \$2{,}712.16$$

two months later; xhere we have used the fact that the margin account paid interest at a nominal rate of 2.7% convertible monthly. Since \$19,120 − \$18,310 = \$810, he got \$2,712.16 − \$810 = \$1,902.16 at settlement. Let i denote Trevor's annual effective yield rate. Trevor held the contract for a sixth of a year, so

$$\$1{,}902.16 = \$2{,}700(1 + i)^{\frac{1}{6}},$$

and therefore, $i \approx -.8777335695 \approx -87.77357\%$.

(3) Rhonda Stallings took a long position on the future contract. As the buyer, she was hoping that the price of the underlier went up, so that she gets a bargain. Rhonda's deposits to the margin account consisted of \$1,485 at the time the position was established, along with \$600 six days later. Since the price at the time she closed her position was \$30,500 − \$29,300 = \$1,200 less than it was at the time she bought the futures contract, the liquidated value of her margin account was \$1,485 + \$600 − \$1,200 = \$885. Her <u>daily</u> yield rate q therefore satisfied the time twelve days equation of value

$$\$1{,}485(1 + q)^{12} + \$600(1 + q)^6 - \$885 = 0,$$

and the quadratic formula gives

$$(1 + q)^6 = \frac{-600 \pm \sqrt{(600)^2 - 4(1{,}485)(-885)}}{2{,}970} = \frac{-600 \pm \sqrt{5{,}616{,}900}}{2{,}970} = \frac{-600 \pm 2,370}{2,970}.$$

Since $(1 + q)^6 \geq 0$, we have

$$(1 + q)^6 = \frac{-600 + 2{,}370}{2,970} = \frac{177}{297}.$$

Therefore, based on a 365 day year, the annual effective interest rate i was

$$i = \left(\frac{177}{297}\right)^{\frac{365}{6}} - 1 \approx (2.116848 \times 10^{-14}) - 1 \approx -100\%.$$

Alternatively, use the **Cashflow worksheet** to compute the yield rate for a six day period; it is about -8.264773801%. Then, convert to the annual effective rate $i \approx (1 - .0824773801)^{\frac{365}{6}} - 1 \approx (2.116848 \times 10^{-14}) - 1 \approx -100\%$. *We note that we would have again received an answer of -100% if we had used a 366 day year.*

(8.7) Price discovery and more kinds of futures

(1) The quote of \$98.75 corresponds to a payoff of \$750,000 + \$2,500(98.75) = \$996,875. At expiration, the three-month rate is 1.55%, so the contract has futures price \$1,000,000 − \$25(155) = \$996,125. Therefore, the price fell by \$996,875 − \$996,125 = \$750. Tuscany, as the buyer, has a position worth −750.

(8.8) Options

(1) Let x denote the dollar price per gallon. If $x \leq 2.35$, the price of the option is worthless [so $f(x) = 0$], and if $x \geq 2.35$, then the value is $\$10{,}000(x - 2.35)$ [so $f(x) = \$10{,}000(x - 2.35) = \$(10{,}000x - 23{,}500)$]. Thus, the graph of $f(x)$ is coincident with the x-axis from $x = 0$ to $x = 2.35$ and for $x \geq 2.35$, it is coincident with the line $y = 10{,}000x - 23{,}500$.

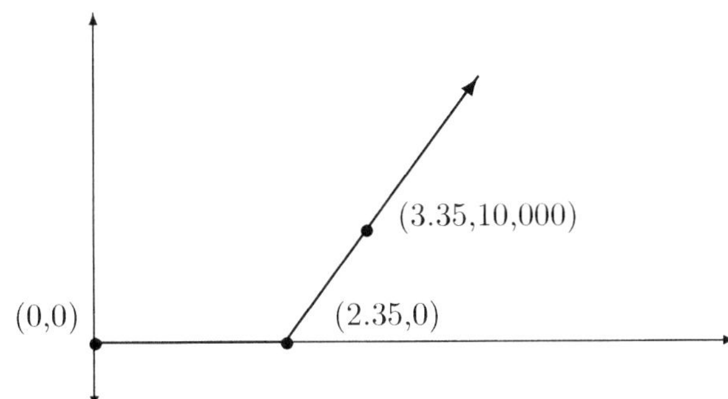

(3) The value of the option at the end of the six months (*half a year*) was 100 × (\$55.80 − \$54) = \$180. Therefore, if we denote Sasha's annual yield rate by i, since Sasha paid \$325.50 for the option, we have

$$\$325.50(1+i)^{\frac{1}{2}} = \$180 \quad \text{and} \quad i = \left(\frac{180}{325.5}\right)^2 - 1 \approx -.694196097 \approx -64.41961\%.$$

We note that this is negative even though the price of the stock went up. If Sasha had purchased stock, then the fact that the price went up would automatically give her a positive yield rate. In fact, her annual yield rate i would have satisfied

$$\$54.25(1+i)^{\frac{1}{2}} = \$55.80,$$

so the yield rate would have been

$$i = \left(\frac{55.8}{54.25}\right)^2 - 1 \approx .057959184 \approx 5.79592\%.$$

(8.9) Using replicating portfolios to price options

(1) We have $S_0 = \$2{,}380$, $K = \$2{,}400$, $u = \$2{,}520$, and $d = \$2{,}220$. Then $V_u = \$2{,}520 - \$2{,}400 = \$120$, and $V_d = \$0$. Hence, by Important Fact (8.9.3), if i is the annual effective risk-free interest rate, we

have

$$\begin{aligned}\$72 = c_0 &= \frac{uV_d - dV_u}{(u-d)(1+i)^T} + \left(\frac{V_u - V_d}{u-d}\right)S_0 \\ &= \frac{(\$2{,}520)(\$0) - (\$2{,}220)(\$120)}{(\$2{,}520 - \$2{,}220)(1+i)^{\frac{1}{4}}} + \left(\frac{\$120 - \$0}{\$2{,}520 - \$2{,}220}\right)(\$2{,}380) \\ &= \frac{-\$888}{(1+i)^{\frac{1}{4}}} + \$952.\end{aligned}$$

Therefore, $i = \left(\frac{888}{880}\right)^4 - 1 \approx 3.686251622\% \approx 3.68625\%$.

(3) The call option expires in three years with strike price $K = \$20{,}000$ and the risk-free interest rate is an annual effective 4%. The initial price of the underlier is $S_0 = \$18{,}400$, the possible underlier prices after one year are

$$S_u = (1.1)(\$18{,}400) = \$20{,}240 \quad \text{and} \quad S_d = (.92)(\$18{,}400) = \$16{,}928,$$

and the possible underlier prices after two years are

$$S_{uu} = (1.1)^2(\$18{,}400) = \$22{,}264,\ S_{ud} = (1.1)(.92)(\$18{,}400) = \$18{,}620.80,$$

and

$$S_{dd} = (.92)^2(\$18{,}400) = \$15{,}573.76.$$

At the end of three years, there are four possible underlier prices, namely

$$S_{uuu} = (1.1)^3(\$18{,}400) = \$24{,}490.40,\ S_{uud} = (1.1)^2(.92)(\$18{,}400) = \$20{,}482.88,$$

$$S_{udd} = (1.1)(.92)^2(\$18{,}400) = \$17{,}131.136,\ S_{ddd} = (.92)^3(\$18{,}400) = \$14{,}327.8592.$$

Note that while you can't actually pay a non-integer number of cents, for no-arbitrage pricing purposes, we do not round to the nearest cent.

The progression of underlier prices is indicated in the following large tree, which has six numbered two-branch subtrees; as usual, we indicate the price of the option below the price of the underlier.

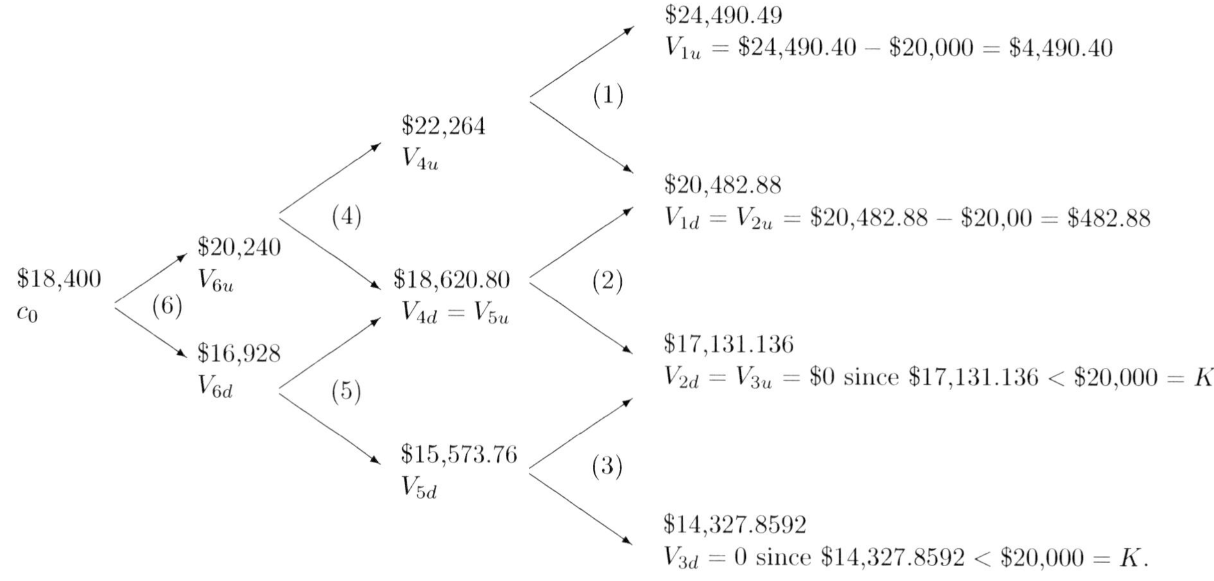

Subtree (1) is based on the price of the underlier having gone up in each of the first two two years, so that we begin with underlier price $S_{uu} = \$22{,}264$. There are two branches, one for the possibility that the price of the underlier goes up again (so that at time three years it is $S_{uuu} = \$24{,}490.40$), the other for the circumstance that the price goes down to $S_{uud} = \$20{,}482.88$. Since the strike price is \$20,000, in the first case, the option will pay $V_{1u} = \$24{,}490.40 - \$20{,}000 = \$4{,}490.40$, while in the second case, it will pay $V_{1d} = \$20{,}482.88 - \$20{,}000 = \$482.88$. Therefore, if you wish to replicate the option payments of this subtree, you should invest an amount f_1 at the riskfree 4% annual effective interest rate and pay $\$22{,}264\Delta_1$ to purchase Δ_1 shares of the underlier where, by Important Fact (8.3.9),

$$f_1 = \frac{S_{uuu}V_{1d} - S_{uud}V_{1u}}{(S_{uuu} - S_{uud})(1+i)} = \frac{(\$24{,}490.40)(\$482.88) - (\$20{,}482.88)(\$4{,}490.40)}{(\$24{,}490.40 - \$20{,}482.88)(1.04)} \approx -\$19{,}230.76923,$$

and

$$\Delta_1 = \frac{V_{1u} - V_{1d}}{S_{uuu} - S_{uud}} = \frac{\$4{,}490.40 - \$482.88}{\$24{,}490.40 - \$20{,}482.88} = 1.$$

Therefore, the no-arbitrage price of the portfolio consisting of the assets purchased at time 2 is

$$f_1 + (\Delta_1)(S_{uu}) = -\$19{,}230.76923 + (\$22{,}264)(1) \approx \$3{,}033.230769.$$

Moreover, since the portfolio replicates the option, by "the law of one price"

$$V_{4u} \approx \$3{,}033.230769.$$

In a similar manner, we consider Subtree (2), the subtree based on the underlier having gone up in one of the two first years and down in the other; the price of the underlier at the beginning of the third year is then \$18,620.80. If the underlier price goes up in the third year, it rises to \$20,482.80 and the value of the option is $V_{2u} = \$20{,}482.88 - \$20{,}000 = \$482.88$. On the other hand, a third year drop in the underlier price will result in the underlier selling for \$17,131.136; since this is below the strike price, the option is worthless ($V_{2d} = 0$). To replicate the option payoffs, invest an amount f_2 at the riskfree 4% annual effective interest rate and pay $\$18{,}620.80\Delta_2$ to purchase Δ_2 shares of the underlier where

$$f_2 = \frac{S_{uud}V_{2d} - S_{udd}V_{2u}}{(S_{uud} - S_{udd})(1+i)} = \frac{(\$20{,}482.88)(\$0) - (\$17{,}131.136)(\$482.88)}{(\$20{,}482.88 - \$17{,}131.136)(1.04)} \approx -\$2{,}373.128205,$$

and

$$\Delta_2 = \frac{V_{2u} - V_{2d}}{S_{uud} - S_{udd}} = \frac{\$482.88 - \$0}{\$20{,}482.88 - \$17{,}131.136} \approx .144068282;$$

again, we have used Important Fact (8.9.3). It follows that the time 2 arbitrage-free option price would be $V_{4d} = V_{5u} = f_2 + (\Delta_2)(S_{ud}) \approx -\$2{,}373.128205 + (.144068282)(\$18{,}620.80) \approx \$309.5384615$.

Likewise,to replicate the option values for subtree (3), the subtree predicated on the underlier having gone down in each of the first two-years so that it is \$15,573.76 at the beginning of the third year, we should invest f_3 and buy Δ_3 shares of the underlier (for $\$15{,}573.76\Delta_3$), where

$$f_3 = \frac{S_{udd}V_{3d} - S_{ddd}V_{3u}}{(S_{udd} - S_{ddd})(1+i)} = \frac{(\$17{,}131.136)(\$0) - (\$14{,}327.8592)(\$0)}{(\$17{,}131.136 - \$14{,}327.8592)(1.04)} = \$0,$$

and

$$\Delta_3 = \frac{V_{3u} - V_{3d}}{S_{udd} - S_{ddd}} = \frac{\$0 - \$0}{\$17{,}131.136 - \$14{,}327.8592} = 0.$$

It follows that $V_{5d} = \$0$.

Having observed the pattern of our calculations, to replicate Subtree (4), the subtree for the second year if the price went up in the first year, we want

$$f_4 = \frac{S_{uu}V_{4d} - S_{ud}V_{4u}}{(S_{uu} - S_{ud})(1+i)} = \frac{(\$22{,}264)(\$309.538461) - (\$18{,}620.80)(\$3{,}033.230769)}{(\$22{,}264 - \$18{,}620.80)(1.04)} \approx -\$13,088.03419,$$

and

$$\Delta_4 = \frac{V_{4u} - V_{4d}}{S_{uu} - S_{ud}} = \frac{\$3{,}033.230769 - \$309.538461}{\$22{,}264 - \$18{,}620.80} \approx .747609878$$

Therefore $V_{6u} = -\$13{,}088.03419 + (.747609878)(\$20{,}240) \approx \$2{,}043.589744$.

For subtree (5), we have

$$f_5 = \frac{S_{ud}V_{5d} - S_{dd}V_{5u}}{(S_{ud} - S_{dd})(1+i)} = \frac{(\$18{,}620.80)(\$0) - (\$15{,}573.76)(\$309.5384615)}{(\$18{,}620.80 - \$15{,}573.76)(1.04)} \approx -\$1{,}521.236029,$$

and

$$\Delta_5 = \frac{V_{5u} - V_{5d}}{S_{uu} - S_{ud}} = \frac{\$309.5384615 - \$0}{\$18{,}620.80 - \$15{,}573.76} \approx .101586609.$$

Therefore $V_{6d} = -\$1{,}521.236029 + (\$16{,}928)(.101586609) \approx \$198.4220907$.

Finally, we look at the subtree for the first year, subtree (6).

$$f_6 = \frac{S_uV_{6d} - S_dV_{6u}}{(S_u - S_d)(1+i)} = \frac{(\$20,240)(\$198.416,928)(\$2{,}043.589744)}{(\$20{,}240 - 16{,}928)(1.04)} \approx -\$8{,}877.341156,$$

and

$$\Delta_6 = \frac{V_{6u} - V_{6d}}{S_u - S_d} = \frac{\$2{,}043.589744 - \$198.4220907}{\$20{,}240 - \$16{,}928} \approx .557115837.$$

Therefore $c_0 = -\$8,877.341156 + (\$18{,}400)(.557115837) \approx \$1,373.590249 \approx \$1{,}373.59$.

(8.10) Using weighted averages to price options: risk-neutral probabilities

(1) The current price of the underlier is $S_0 = 1{,}000 \times \$64 = \$6{,}400$, and the strike price K is also \$6,400. The risk-free interest rate for each quarter is a nominal 6% convertible quarterly, so the quarterly risk-free interest rate is $\frac{6\%}{4} = 1.5\%$. The stock price at the end of one quarter is either $u_1 = (1 + .07)(\$6{,}400) = \$6{,}848$ or $d_1 = (1 - .013)(\$6{,}400) = \$5{,}568$. Therefore, using Equation (8.10.1), we find that the risk-neutral probability of an increase in the first quarter is

$$p_1^* = \frac{(\$6{,}400)(1.015) - \$5{,}568}{\$6{,}848 - \$5{,}568} = .725.$$

According to Fact (8.10.10), since the quarters are taken to be of equal length, the probability p_2^* of an increase in the second quarter is also equal to .725;

$$p_2^* = p_1^* = .725.$$

If the stock price goes up by 7% each quarter, so that the underlier price after two quarters is $(1 + .07)^2(\$6{,}400) = \$7{,}327.36$, the option will be exercised; it's value is $\$7{,}327.36 - \$6{,}400 = \$927.36$.

On the other hand, if the price goes down by 13% in at least one the two periods, the stock price will be below \$6,400, and the option has no value. Therefore,

$$\begin{aligned} c_0 &= p_1^* p_2^* (\$927.36)(1.015)^{-2} + [(1-p_1^*)p_2^* + p_1^*(1-p_2^*)] (\$0)(1.015)^{-2} + (1-p_1*)(1-p_2^*)(\$0)(1.015)^{-2} \\ &= p_1^* p_2^* (\$927.36)(1.015)^{-2} \\ &= (.725)^2 (\$927.36)(1.015)^{-2} \\ &\approx \$473.1428571, \end{aligned}$$

and the no-arbitrage price of the call option is \$473.14.

(3) As reported in our solution to Problem (8.9.1), we have $S_0 = \$2{,}380$, $K = \$2{,}400$, $u = \$2{,}520$, $d = \$2{,}220$, $V_u = \$2{,}520 - \$2{,}400 = \$120$, and $V_d = \$0$. Moreover, $c_0 = \$72$, and the problem concerns a time period of two quarters, considered as a sequence of two periods of length $T = \frac{1}{4}$. Letting p^* denote the risk-neutral probability for one quarter and using i to represent the annual effective risk-free interest rate, Equation (8.10.2) thus gives us

$$\begin{aligned} \$72 &= \left[V_u(1+i)^{-T}\right] p^* + \left[V_d(1+i)^{-T}\right](1-p^*) \\ &= \left[\$120(1+i)^{-\frac{1}{4}}\right] p^* + \left[\$0(1+i)^{-\frac{1}{4}}\right](1-p^*) \\ &= \left[\$120(1+i)^{-\frac{1}{4}}\right] p^*. \end{aligned}$$

It follows that

$$(1+i)^{\frac{1}{4}} p^* = \frac{72}{120} = .6.$$

On the other hand, according to Fact (8.10.5),

$$\begin{aligned} \$2{,}380 = S_0 &= p^*\left[u(1+i)^{-T}\right] + (1-p^*)\left[d(1+i)^{-T}\right] \\ &= p^*\left[\$2{,}520(1+i)^{-\frac{1}{4}}\right] + (1-p^*)\left[\$2{,}220(1+i)^{-\frac{1}{4}}\right] \\ &= p^*(1+i)^{-\frac{1}{4}}(\$2{,}520 - \$2{,}220) + \$2{,}220(1+i)^{-\frac{1}{4}} \\ &= .6(\$2{,}520 - \$2{,}220) + \$2{,}220(1+i)^{-\frac{1}{4}} \\ &= \$180 + \$2{,}220(1+i)^{-\frac{1}{4}}. \end{aligned}$$

Therefore,

$$(1+i)^{\frac{1}{4}} = \frac{2{,}220}{2{,}200} \quad \text{and} \quad i = \left(\frac{2{,}220}{2{,}200}\right)^4 \approx 3.686251622\% \approx 3.68625\%$$

as in problem (8.9.1).

(5) The call option expires in three years with strike price $K = \$20{,}000$, and the risk-free interest rate is an annual effective 4%. The initial price of the underlier is $S_0 = \$18{,}400$, the possible underlier prices after one year are $S_u = (1.1)(\$18{,}400) = \$20{,}240$ and $S_d = (.92)(\$18{,}400) = \$16{,}928$. Therefore, by Equation (8.10.1), the risk-neutral probability for a one-year period is

$$p^* = \frac{\$18{,}400(1.04) - \$16{,}928}{20{,}240 - 16{,}928} = \frac{2{,}208}{3{,}312} = \frac{2}{3}.$$

We next consider the price of the underlier at the end of three years.

- The probability that the underlier price goes up in all three years, so that it is equal to

$$S_{uuu} = (1.1)^3(\$18{,}400) = \$24{,}490.40,$$

is

$$(p^*)^3 = \left(\frac{2}{3}\right)^3 = \frac{8}{27}.$$

- The risk-neutral probability that the underlier goes up in exactly two of the three years, so that it is equal to

$$S_{uud} = (1.1)^2(.92) = (\$18{,}400) = \$20{,}482.88,$$

is

$$3(p^*)^2(1-p^*) = 3\left(\frac{2}{3}\right)^2\left(\frac{1}{3}\right) = \frac{12}{27}.$$

- The risk-neutral probability that that there is a unique year in which the price goes up, so that it is equal to

$$S_{udd} = (1.1)(.92)^2(\$18{,}400) = \$17{,}131.136,$$

is

$$3(1-p^*)(1-p^*)^2 3\left(\frac{1}{3}\right)^2\left(\frac{2}{3}\right) = \frac{6}{27}.$$

- The risk-neutral probability that the option goes down an all three years , ending at

$$S_{ddd} = (.92)^3(\$18{,}400) = \$14{,}327.8592,$$

is

$$(1-p^*)^3 = \left(\frac{1}{3}\right)^3 = \frac{1}{27}.$$

Since the strike price is $K = \$20{,}000$, the option's value at the end of three years is \$4,490.40, \$482.88, or \$0, and these occur with probabilitites $\frac{8}{27}$, $\frac{12}{27}$, and $\frac{6}{27} + \frac{1}{27} = \frac{7}{27}$, respectively.

The arbitrage-free option price is the expected value of the present value of the option pay-off; it is

$$c_0 = \left(\frac{8}{27}\right)\left[(\$4{,}490.40)(1.04)^{-3}\right] + \left(\frac{12}{27}\right)\left[(\$482.88)(1.04)^{-3}\right] + \left(\frac{7}{27}\right)\left[(\$0)(1.04)^{-3}\right]$$
$$\approx \$1{,}373.590249 \approx \$1{,}373.59.$$

The \$1,373.59 arbitrage-free option price was also calculated in the solution to Problem (8.9.3), but our calculation here was considerably shorter than the one in that solution.

(8.11) Swaps

(1) Great Savings Bank is due to receive $\$35{,}000{,}000 \times \frac{.0525}{4} = \$459{,}375$ each quarter, but since cashflows are made on a netted basis, each quarter Great Savings Bank's liability must be deducted. These quarterly liabilities are figured using the LIBOR and an "actual/360" basis, so we must determine exact day counts for each quarter; exact day counts for the four quarters ending on July 1, 1994, October 1, 1994, January 1, 1995, and April 1, 1995 are $30 + 31 + 30 = 91$, $31 + 31 + 30 = 92$, $31 + 30 + 31 = 92$, $31 + 28 + 31 = 90$, respectively Noting the LIBOR rates quoted in the statement of the problem, GSB's liabilities are calculated to be

$$\left(\frac{91}{360}\right)(.0425)(\$35{,}000{,}000) \approx \$376{,}006.94 \quad \text{on July 1, 1994,}$$
$$\left(\frac{92}{360}\right)(.04875)(\$35{,}000{,}000) \approx \$436{,}041.67 \quad \text{on October 1, 1994,}$$
$$\left(\frac{92}{360}\right)(.05688)(\$35{,}000{,}000) \approx \$508{,}760 \quad \text{on January 1, 1995,}$$
$$\left(\frac{90}{360}\right)(.06328)(\$35{,}000{,}000) \approx \$553{,}700 \quad \text{on April 1, 1995.}$$

The first two of these liabilities are less than \$459,375 (*the amount GSB is due to receive at the end of each quarter*) so GSB receives

$$\$459{,}375 - \$376{,}006.94 = \$83{,}368.06 \quad \text{on July 1, 1994}$$

and

$$\$459{,}375 - \$436{,}041.67 = \$23{,}333.33 \quad \text{on October 1, 1994.}$$

On the other hand, at the next two dates, the liabilities exceed the \$459,375 due, so GSB must make a payment; the amounts of these outflows are

$$\$508{,}760 - \$459{,}375 = \$49{,}385 \quad \text{on January 1, 1995,}$$

and

$$\$553{,}700 - \$459{,}375 = \$94{,}325 \quad \text{on April 1, 1995.}$$

(3) The initial equity price is \$56,400, and this amount will be used to calculate the semi-annual payments Mrs. Markov is due, while the returns on the stock dictate the amunts she is liable for. More specifically, if not for netting, Mrs. Markov would be entitled to receive semiannual payments for $(\$56{,}400)\left(\frac{.025}{2}\right) =$ \$750, but she would also be obligated to pay the semi-annual dividends and price appreciation, namely

$$\$150 + (\$57{,}600 - \$56{,}400) = \$1{,}350 \text{ at the end of six months,}$$
$$\$0 + (\$57{,}900 - \$57{,}600) = \$300 \text{ at the end of twelve months,}$$
$$\$0 + (\$55{,}500 - \$57{,}900) = -\$2{,}400 \text{ at the end of eighteen months,}$$
$$\$0 + (\$54{,}400 - \$55{,}500) = -\$1{,}500 \text{ at the end of twenty-four months.}$$

Therefore, using netting,

at the end of six months, Mrs Markov pays \$1,350 − \$705 = \$645,
at the end of twelve months, Mrs Markov receives \$705 − \$300 = \$405
at the end of eighteen months, Mrs Markov receives \$705 − (−\$2,400) = \$3,105
at the end of twenty-four months, Mrs Markov receives \$705 − (−\$1,500) = \$2,205.

Chapter 8 review problems

(1) The bond we are to price has $F = C =$ \$1,000, $m = 4$, $\alpha = 5\%$, $r = \frac{5\%}{4} = 1.25\%$, and coupon amount $Fr =$ (\$1,000)(.0125) = \$12.50. The bond hence provides cashflows of \$12.50 at the end of the first quarter, \$12.50 at the end of the second quarter, \$12.50 at the end of the third quarter, and

\$12.50 + \$1,000 = \$1,012.50 at the end of the fourth quarter; based on the yields for zero coupon bonds, in order to avoid creating an arbitrage opportunity, these cashflows should be discounted using the annual effective rates 2.25%, 2.45%, 2.95%, and 3.35% respectively. So, the no-arbitrage price of the bond is

$$P = \frac{\$12.50}{(1.0225)^{\frac{1}{4}}} + \frac{\$12.50}{(1.0245)^{\frac{2}{4}}} + \frac{\$12.50}{(1.0295)^{\frac{3}{4}}} + \frac{\$1{,}012.50}{1.0335} \approx \$1{,}016.691377 \approx \$1{,}016.69.$$

(3) First suppose that $X > \$4{,}200 = K$. Then in six months, the option is sure to have value and hence, we presume, to be exercised. Thus, you now have two ways to assure you hold the underlier in six months; either you buy the underlier now, at a cost of $S_0 = \$4{,}238$ or you buy the option for c_0 and invest $K(1+i)^{-\frac{1}{2}} = \$4{,}200(1.058)^{-\frac{1}{2}}$, then exercise the option (to buy the underlier) in six months, using the \$4,200 accumulated value of your investment. The "law of one price" says that if c_0 was determined using no-arbitrage pricing, then these two methods must cost you the same amount, and therefore

$$\$4{,}238 = c_0 + \$4{,}200(1.058)^{-\frac{1}{2}}.$$

Solving for c_0, we find $c_0 = \$4{,}238 - \$4{,}200(1.058)^{-\frac{1}{2}} \approx \$154.7454324 \approx \$154.75$. This is not consistent with the option price (figured using no-arbitrage pricing) being \$173.51, so a no-arbitrage price of \$173.51 must be for an underlier with $X \leq \$4{,}200$.

Now, suppose that, as required by no-arbitrage pricing, $c_0 = \$173.51$. We then have $S_0 = \$4{,}238$, $K = \$4{,}200$, and the underlier price either goes up to $u = \$4{,}424$, or down to $d = X \leq \$4{,}200$. So, $V_u = \$4{,}424 - \$4{,}200 = \$224$, and $V_d = \$0$. Let p^* denote the risk-neutral probability for the six month period. Equation (8.10.2) gives us

$$\$173.51 = \$224(1.058)^{-\frac{1}{2}}p^* + \$0(1.058)^{-\frac{1}{2}}(1-p^*) = \$224(1.058)^{-\frac{1}{2}}p^*,$$

and this is equivalent to

$$p^* = \frac{(1.058)^{\frac{1}{2}}(173.51)}{224}.$$

But, according to Equation (8.10.1),

$$p^* = \frac{(\$4{,}238)(1.058)^{\frac{1}{2}} - X}{\$4{,}424 - X}.$$

It follows that

$$\frac{(1.058)^{\frac{1}{2}}(173.51)}{224} = \frac{(\$4{,}238)(1.058)^{\frac{1}{2}} - X}{\$4{,}424 - X}.$$

Therefore

$$(\$4{,}424 - X)\left(\frac{(1.058)^{\frac{1}{2}}(173.51)}{224}\right) = \$4{,}238(1.058)^{\frac{1}{2}} - X,$$

$$X\left[1 - \left(\frac{(1.058)^{\frac{1}{2}}173.51}{224}\right)\right] = \$4{,}238(1.058)^{\frac{1}{2}} - \left(\frac{(\$4{,}424)(1.058)^{\frac{1}{2}}(173.51)}{224}\right)$$

and

$$X = \left[\$4{,}238(1.058)^{\frac{1}{2}} - \left(\frac{(\$4{,}424)(1.058)^{\frac{1}{2}}(173.51)}{224}\right)\right] \Big/ \left[1 - \left(\frac{(1.058)^{\frac{1}{2}}(173.51)}{224}\right)\right]$$

$$\approx \$4{,}105.040138 \approx \$4{,}105.04.$$

Thus, a no-arbitrage option price of \$173.51 requires that the second possible price X is \$4,105.04

Now, let us consider a no-arbitrage price of \$154.75. We have already noted that the equation $c_0 = \$154.75$ occurs whenever $X > \$4{,}200$. Thus, a unique X cannot be determined.

(5) Since the underlier is 100 shares of the stock, $S_0 = \$2{,}400$. Every three months, the price either rises by 7% or falls by 5%; at the end of three months (*the first period*), the stock price is

$$S_u = (1 + .07)(\$2{,}400) = \$2{,}568 \quad \text{or} \quad S_d = (1 - .05)(\$2{,}400) = \$2{,}280,$$

and the possible prices after six months (*two periods*) are

$$S_{uu} = (1 + .07)^2(\$2{,}400) = \$2{,}747.76,\ S_{ud} = (1 + .07)(1 - .05)(\$2{,}400) = \$2{,}439.60,$$
$$\text{and} \quad S_{dd} = (1 - .05)^2(\$2{,}400) = \$2{,}166.$$

We are given a strike price of \$2,300.

Call Option: Now suppose the option is a call option; that is, the holder has the right to buy the underlier at the end of the six months for \$2,300. We display the possible underlier prices in a tree, as usual noting the value of the option below the underlier price.

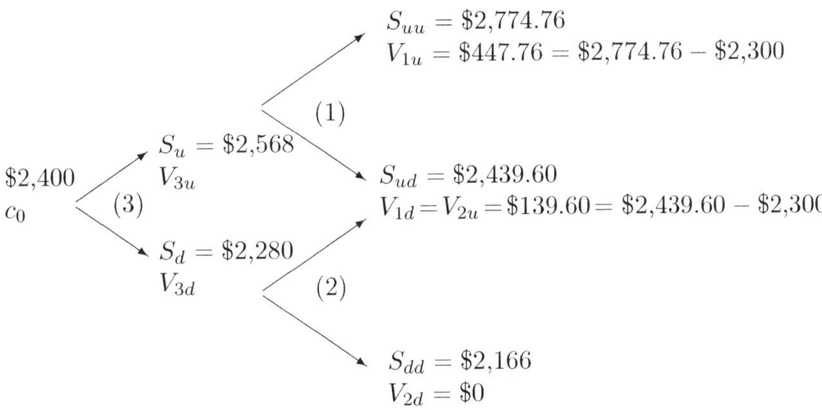

While the price of the call option is most easily calculated using the risk-neutral probability, since we are asked show how dynamic portfolio allocation may be used to replicate the option, we need to follow through the binomial pricing model.

For subtree (1), we want to invest f_1 in the $i = 3.2\%$ annual effective risk-free account and purchase Δ_1 shares of the underlier ($100\Delta_1$ shares of the stock) by paying $\$2{,}1568\Delta_1$. According to Important Fact (8.9.3),

$$\Delta_1 = \frac{V_{1u} - V_{1d}}{S_{uu} - S_{ud}} = \frac{\$447.76 - \$139.60}{\$2{,}774.76 - \$2{,}439.60} = 1,$$

and

$$f_1 = \frac{S_{uu}V_{1d} - S_{ud}V_{1u}}{(S_{uu} - S_{ud})(1+i)^{\frac{1}{4}}} = \frac{(\$2{,}747.76)(\$139.60) - (\$2{,}439.60)(\$447.76)}{(\$2{,}747.76 - \$2{,}439.60)(1.032)^{\frac{1}{4}}} \approx -\$2{,}281.959392.$$

Moreover, again according to Important Fact (8.9.3),

$$V_{3u} = f_1 + \Delta_1 S_u \approx -\$2{,}281.959392 + 1(\$2{,}568) \approx \$286.0406084.$$

Just as we did for subtree (1), we may apply Important Fact (8.9.3) to subtree (2). We should invest f_2 in the $i = 3.2\%$ annual effective risk-free account and purchase Δ_2 shares of the underlier by paying $\$2{,}1280\Delta_2$ where

$$\Delta_2 = \frac{V_{2u} - V_{2d}}{S_{ud} - S_{dd}} = \frac{\$139.60 - \$0}{\$2{,}439.60 - \$2{,}166} \approx .510233918,$$

and

$$f_2 = \frac{S_{ud}V_{2d} - S_{dd}V_{2u}}{(S_{ud} - S_{dd})(1+i)^{\frac{1}{4}}} = \frac{(\$2{,}439.60)(\$0) - (\$2{,}166)(\$139.60)}{(\$2{,}439.60 - \$2{,}166)(1.032)^{\frac{1}{4}}} \approx -\$1{,}096.498024.$$

Furthermore,

$$V_{3d} = f_2 + \Delta_2 S_{ud} \approx -\$1{,}096.498024 + (.510233918)(\$2{,}280) \approx \$66.83530975.$$

Turning to subtree (3), we favor investing f_3 in the $i = 3.2\%$ annual effective risk-free account and purchase Δ_3 shares of the underlier by paying $\$2{,}400\Delta_3$ where

$$\Delta_3 = \frac{V_{3u} - V_{3d}}{S_u - S_d} \approx \frac{\$286.0406084 - \$66.83530975}{\$2{,}568 - \$2{,}280} \approx .761129509,$$

and

$$f_3 = \frac{S_u V_{3d} - S_d V_{3u}}{(S_u - S_d)(1+i)^{\frac{1}{4}}} = \frac{(\$2{,}568)(\$66.83530975) - (\$2{,}280)(\$286.0406084)}{(\$2{,}568 - \$2{,}280)(1.032)^{\frac{1}{4}}} \approx -\$1{,}655.452373.$$

Then,

$$c_o = f_3 + \Delta_3 S_0 \approx -\$1{,}655.452373 + (.761129509)(\$2{,}400) \approx \$171.2584493 \approx \$171.26.$$

So, the cost of the Call option, based on no-arbitrage pricing, is \$171.26.

Now let us turn to *dynamic portfolio allocation* in which we replicate the effect of a call option without actually purchasing the option. We would like to purchase $100\Delta_3 \approx (100)(.761129509) \approx 76$ shares of the stock. So, since the price for a share of stock is $\$2{,}400/100 = \$24$, we wish to spend $(76)(\$24) = \$1{,}824$ to purchase 76 shares of the stock; we assume only integer number of shares can be purchased. We only use \$171.26 of our own money for this purchase, borrowing the remaining $\$1{,}824 - \$171.26 = \$1{,}652.74$ at the 3.2% annual effective interest rate.

At the end of three months, our portfolio consists of 76 shares of the stock, along with a liability of $\$1{,}652.74(1.032)^{\frac{1}{4}}$, and it is time to adjust our holdings; the manner we do this is based on whether the current price of a share of the stock is \$25.68 or \$22.80.

If the price of the stock has risen to \$25.68, we wish to increase the number of shares held to be $100\Delta_1 = 100$ shares, so we purchase an additional $100 - 76 = 24$ shares; we do this using $24 \times \$25.68 = \$616.32$ of newly-borrowed money. After the purchase, we hold 100 shares along with liabilities totaling $\$1{,}652.74(1.032)^{\frac{1}{4}} + \$616.32 \approx \$2{,}282.126155$. Still under the assumption that the stock went up during the first quarter, at the end of another quarter the 100 shares of stock we hold will be worth either \$2,774.76 or \$2,439.60, depending on whether the stock price goes up or down during the second quarter. Since our liabilities will now have grown to $\$1{,}652.74(1.032)^{\frac{1}{2}} + \$616.32(1.032)^{\frac{1}{4}} \approx \$2{,}300.168082 \approx \$2{,}300.17$, after paying off our liability we will have a surplus of either $\$2{,}774.76 - \$2{,}300.17 = \$474.59$ or $\$2{,}439.60 - \$2{,}300.17 = \$139.47$. In each case, the surplus is very close to what the value of the call option would be.

We still need to consider the case where the share price of the stock fell to \$22.80 after one quarter. In this case, the number of shares of the stock you wish to hold is just $100\Delta_2 \approx 51$ shares, so you sell $76 - 51 = 25$ shares, thereby receiving $25 \times \$22.80 = \$570$ which you use to reduce your liability to $\$1{,}652.74(1.032)^{\frac{1}{4}} - \$570 \approx \$1{,}095.806155$. Three months later, this liability has grown to $\left[\$1{,}652.74(1.032)^{\frac{1}{4}} - \$570\right](1.032)^{\frac{1}{4}} \approx \$1{,}104.47$, and your 51 shares of stock have a market value of either $.51 \times \$2{,}439.60 \approx \$1{,}244.20$ or $.51 \times \$2{,}166 \approx \$1{,}104.66$. So, after paying off your loan, you have a surplus of either $\$1{,}244.20 - \$1{,}104.47 = \$139.73$ or $\$1{,}104.66 - \$1,104.47 = \$.19$. Again, we have come very close to replicating the value of the call option at the end of two quarters.

Put Option: Next consider what happens if the option is a put option. We still have the same tree of underlier prices, but the option values look different since the option only has value if the underlier price is *below* the \$2,300 strike price. More specifically, we have

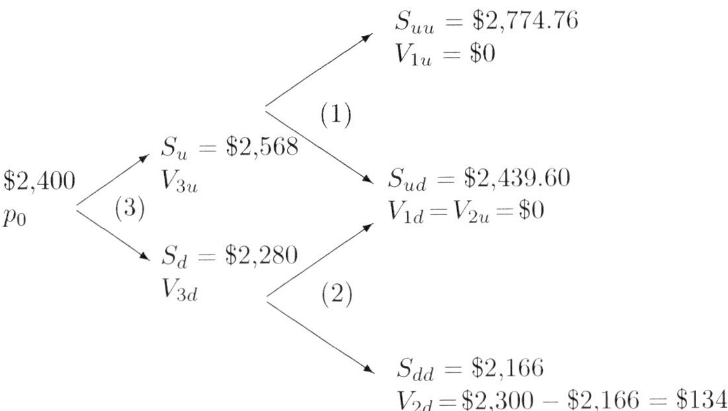

Now, with notation as before except that we use p_0 for the value of our put option at time 0, by Important Fact (8.9.3), we have

for subtree (1):

$$\Delta_1 = \frac{V_{1u} - V_{1d}}{S_{uu} - S_{ud}} = \frac{\$0 - \$0}{\$2{,}774.76 - \$2{,}439.60} = 0;$$

$$f_1 = \frac{S_{uu}V_{1d} - S_{ud}V_{1u}}{(S_{uu} - S_{ud})(1+i)^{\frac{1}{4}}} = \frac{(\$2{,}747.76)(\$0) - (\$2{,}439.60)(\$0)}{(\$2{,}747.76 - \$2{,}439.60)(1.032)^{\frac{1}{4}}} = 0;$$

$$V_{3u} = f_1 + \Delta_1 S_u = \$0 + (0)(\$2{,}568) = \$0.$$

for subtree (2):

$$\Delta_2 = \frac{V_{2u} - V_{2d}}{S_{ud} - S_{dd}} = \frac{\$0 - \$134}{\$2{,}439.60 - \$2{,}166} \approx -.489766082;$$

$$f_2 = \frac{S_{ud}V_{2d} - S_{dd}V_{2u}}{(S_{ud} - S_{dd})(1+i)^{\frac{1}{4}}} = \frac{(\$2{,}439.60)(\$134) - (\$2{,}166)(\$0)}{(\$2{,}439.60 - \$2{,}166)(1.032)^{\frac{1}{4}}} \approx \$1{,}185.461368;$$

$$V_{3d} = f_2 + \Delta_2 S_{ud} \approx \$1{,}185.461368 + (-.489766082)(\$2{,}280) \approx \$68.79470131;$$

for subtree (3):

$$\Delta_3 = \frac{V_{3u} - V_{3d}}{S_u - S_d} \approx \frac{\$0 - \$68.79470131}{\$2{,}568 - \$2{,}280} \approx -.238870491;$$

$$f_3 = \frac{S_u V_{3d} - S_d V_{3u}}{(S_u - S_d)(1+i)^{\frac{1}{4}}} = \frac{(\$2{,}568)(\$668.79470131) - (\$2{,}280)(\$0)}{(\$2{,}568 - \$2{,}280)(1.032)^{\frac{1}{4}}} \approx \$608.6079158;$$

$$p_o = f_3 + \Delta_3 S_0 \approx \$608.6079158 + (-.238870491)(\$2{,}400) \approx \$35.31873827 \approx \$35.32.$$

The \$35.52 price may also be found by using risk-neutral probabilities. Moreover, it may be easily checked using put-call parity and a \$171.26 price for the call option; this is because Equation (8.8.4) gives

$$p_0 + \$2{,}400 = \$2{,}300(1.032)^{-\frac{1}{2}} + c_0,$$

and

$$\left[\$2{,}300(1.032)^{-\frac{1}{2}} + \$171.72\right] - \$2{,}400 \approx \$35.32028901 \approx \$35.32.$$

However, as was the case for the call option, the values we found for the Δ_1, Δ_2, Δ_3, f_1, f_2, and f_3 are needed for dynamic portfolio allocation, the replicating of the option without actually purchasing it.

We now turn to the dynamic portfolio allocation. At the beginning of the two-quarter interval, sell $-100\Delta_3 \approx -(100)(-.238870491) \approx 24$ shares of the stock. (*Recall* [see Section (7.4)] *that you do not need to own shares in order to sell them using a short sale.*) Since the initial price of the stock is \$24 per share, this will provide an inflow of 24 × \$24 = \$576. Since we are also willing to use \$35.32 of our own money, we have \$576 + \$35.32 = \$611.32 to invest at $i = 3.2\%$. Three months later, this will have grown to $\$611.32(1.032)^{\frac{1}{4}} \approx \$616.1529452$.

Three months later, if the price of the share has gone up to \$25.68, then we want to hold exactly $100\Delta_1 = 0$ shares. Since we currently have sold 24 shares, we purchase 24 shares to offset these. This costs 24 × \$25..68 = \$616.32, of which \$.17 needs to be borrowed. Three months later, the price of the stock is of no relevance to the value of our portfolio since we hold no stock, and since $.17(1.032)^{\frac{1}{4}} \approx \$.171343978$, you are about \$.17 behind where you would be had you used the \$35.32 to purchase the put option.

If three months after the sale of 24 shares, the price of the share has gone down to \$22.80, then the number of shares you want is $100\Delta_2 \approx 100(-.489766082) \approx -49$ shares, so you need to sell another 25 shares. This brings in 25×\$22.80 = \$570. Combined with the accumulated value of the \$611.32 you held three months previously, you have about $\$611.32(1.032)^{\frac{1}{4}} + \$570 \approx \$1{,}186.152945$, and three months later this has grown by a factor of $(1.032)^{\frac{1}{4}}$ to about \$1,195.53. At this point, you buy back the forty-nine shares for either (.49)(\$2,439.60) ≈ \$1,195.40 or for (.49)(\$2,166) = \$1,061.34. In the first case, you end up with \$1,195.53 − \$1,195.40 = \$.13 , while in the latter you have \$1,195.53 − \$1,061.34 = \$134.19.

In all cases, the value of our portfolio is within a few cents of the value of the put option, and these small discrepancies are due to our insistence that we rounded to the nearest integer number of shares.

(7) Marjorie bought the futures contract on January 5th for \$32,460 and made the requisite \$2,500 margin deposit. Further deposits will only come when she falls below the maintenance margin requirement, and if she is required to make a deposit, she will add sufficient money to bring her balance back up to \$2,500. We note that as the buyer of the futures contract, having locked in a price, her balance gains if the price goes up and falls if it goes down.

The price used for marking-to-market on the purchase date was \$32,075. Since \$32,460 − \$32,075 = \$385, Marjories's account balance is adjusted downward to \$2,500 − \$385 = \$2,115.

On January 6, the marking-to-market price is \$31,200. This is a drop of \$32,075 − \$31,200 = \$875, so Marjorie's account balance drops to \$2,115 − \$875 = \$1,240. Since this is below \$2,000, she must make a deposit. The deposit amount is \$2,500 − \$1,240 = \$1,260, bringing her balance back up to \$2,500.

On January 7, the price of the futures contract has decreased by another \$1,400 to \$29,800, and this results in a drop in Marjorie's account balance to \$1,100. To bring it back up to \$2,000, she deposits \$1,400.

Thereafter, the price of the futures contract rises, and there is a corresponding rise in the balance in Marjorie's account. Since \$30,600 − \$29,800 = \$800, on January 8 her balance is \$2,500 + \$800 = \$3,300. The January 9th marking-to-market value is \$32,975 − \$30,600 = \$2,375 higher than the January 8th value, so Marjorie's balance increases to \$3,300 + \$2,375 = \$5,675. The January 12th sale price is

\$31,800, which is \$32,975 − \$31,800 = \$1,175 less than the last (January 9th) price used for marking-to-market, so Marjorie gets \$5,675 − \$1,175 = \$4,500 when she sells the contract. Note that the January 12th marking-to-market price is not needed.

As requested in the problem (*and given in the answer section of the text, page 483*), these results may be presented in a chart:

Time	Balance	Deposit	New Balance
1/5 at purchase	\$0	\$2,500	\$2,500
1/5	\$2,115	\$0	\$2,115
1/6	\$1,240	\$1,260	\$2,500
1/7	\$1,100	\$1,400	\$2,500
1/8	\$3,300	\$0	\$3,300
1/9	\$5,675	\$0	\$5,675
1/12 at sale	\$4,500	−\$4,500	\$0

C H A P T E R 9

Interest rate sensitivity

(9.1) Overview

(1) Let F denote the face value of the one-year 6% bond. With notation as usual, this bond has $m = 2$ coupons per year, a nominal coupon rate $\alpha = 6\%$, and a coupon rate $r = \frac{\alpha}{2} = 3\%$. At the end of one year, the holder of this bond receives the face amount F (*since the bond is a par value bond*), along with the coupon amount $Fr = F(.03)$. So, at the end of one year, the holder receives amounts totaling $(1.03)F$, and we need this to be equal to \$10,000, the amount of his liability at that time. Therefore, we choose $F = \$9{,}708.74$. [*Note that* $(\$9{,}708.74)(1.03) = \$10{,}000.002$.] Since $Fr = (\$9{,}708.74)(.03) = \$291.2622$, his bond pays \$291.26 after six months as well as \$10,000 at the end of the year. Noting that $\$291.26(1.02)^{-1} + \$10{,}000(1.02)^{-2} \approx \$9{,}897.236832 \approx \$9{,}897.24$, pricing based on a 4% nominal yield convertible semiannually results in this one-year bond costing \$9,897.24.

We have yet to consider Leland's \$5,000 liability; this is an obligation to pay \$5,000 in six months, and so far, the only inflow due at that time is a \$291.26 coupon payment. Thus, Leland should invest an amount F' in the six-month bond, so as to have a redemption payment of $\$5{,}000 - \$291.26 = \$4{,}708.74$ in six months. Since the six-month zero-coupon has a 3.0225% annual yield, we look for F' with $F' = \$4{,}708.74(1.030225)^{-\frac{1}{2}} \approx \$4{,}639.152709 \approx \$4{,}639.15$. We take $F' = \$4{,}635.16$, and since $\$4{,}639.15(1.030225)^{\frac{1}{2}} \approx \$4{,}708.73725$, the six month bond has the desired \$4,708.74 redemption amount.

Leland's total bond expenditure is $\$9{,}897.24 + \$4{,}639.15 = \$14{,}536.39$.

(9.2) Macaulay and modified duration

(1) Since $F = \$1{,}000$, $r = 6\%$, and $C = \$1{,}200$, the bond pays \$60 at times $t = 1, 2, \ldots, 9$ and a total of $\$60 + \$1{,}200 = \$1{,}260$ at $t = 10$. Therefore, the price to yield the buyer 8% is

$$P(8\%) = \$60 a_{\overline{10}|8\%} + \$1{,}200(1.08)^{-10} \approx \$958.437069.$$

Moreover, using Important Fact (9.2.21), the Macaulay duration $D(8\%, \infty)$ is

$$\begin{aligned} D(8\%, \infty) &= \left(\sum_{k=1}^{9} \left[\frac{\$60(1.08)^{-k}}{P(8\%)} \right] k \right) + \left[\frac{\$1{,}260(1.08)^{-10}}{P(8\%)} \right] (10) \\ &= \frac{\$60}{P(8\%)} (Ia)_{\overline{9}|8\%} + \$12{,}600 \left[\frac{(1.08)^{-10}}{P(8\%)} \right] \\ &\approx \frac{\$60}{\$958.437069} (Ia)_{\overline{9}|8\%} + \$12{,}600 \left[\frac{(1.08)^{-10}}{\$958.437069} \right] \\ &\approx 1.756295466 + 6.089328277 \\ &\approx 7.845623743 \approx 7.84562. \end{aligned}$$

If you prefer to compute $D(8\%, 1)$ directly from the definition of the modified duration [Equation (9.2.12)] and then use Equation (9.2.18) to determine the Macaulay Duration $D(8\%, \infty)$, you will need to calculate $P'(8\%)$. Note that

$$P(i) = \left[\$60 \sum_{k=1}^{10} (1+i)^{-k}\right] + \$1{,}200(1+i)^{-10},$$

and

$$\begin{aligned} P'(i) &= \left[\$60 \sum_{k=1}^{10} -k(1+i)^{-k-1}\right] - \frac{\$12{,}000}{(1+i)^{11}} \\ &= \$60\left[-\frac{1}{(1+i)^2} - \frac{2}{(1+i)^3} - \cdots - \frac{10}{(1+i)^{11}}\right] - \frac{\$12{,}000}{(1+i)^{11}} \\ &= -\frac{\$60}{(1+i)}\left[\frac{1}{(1+i)^1} + \frac{2}{(1+i)^2} + \cdots + \frac{10}{(1+i)^{10}}\right] - \frac{\$12{,}000}{(1+i)^{11}} \\ &= -\frac{\$60}{(1+i)}(Ia)_{\overline{10}|i} - \frac{\$12{,}000}{(1+i)^{11}}. \end{aligned}$$

In particular,

$$\begin{aligned} P'(8\%) &= -\frac{\$60}{(1.08)}(Ia)_{\overline{10}|8\%} - \frac{\$12{,}000}{(1.08)^{11}} \\ &\approx \left[-\frac{\$60}{(1.08)}(32.68691288) + \$5{,}146.594312\right] \\ &\approx \$6{,}962.533916. \end{aligned}$$

Thus,

$$D(8\%, 1) = -\frac{P'(8\%}{P(8\%)} \approx -\frac{\$6,962.533916}{\$958.437069} \approx 7.264466429,$$

and

$$D(8\%, \infty) = D(8\%, 1)(1.08) \approx (7.264466429)(1.08) \approx 7.845623743 \approx 7.84562.$$

(3) Since the bond has a single payment time, the Macaulay duration is just equal to that time;

$$D(.05, \infty) = 8.$$

But according to Equation (9.2.18),

$$D(.05, \infty) = D(.05, 1)(1.05).$$

Thus,

$$D(.05, 1) = \frac{8}{1.05} \approx 7.61905.$$

(5) The interest rate for a half-year period is $(1+i)^{\frac{1}{2}} - 1$, and \$50 dividends are paid semiannually, with the next dividend to be paid in exactly half a year. These dividends form a perpetuity, and the present value of this perpetuity is the price of the stock. Thus, the price of the stock is given as a function of the annual effective interest rate i by

$$P(i) = \frac{\$50}{(1+i)^{\frac{1}{2}} - 1} = \$50\left[(1+i)^{\frac{1}{2}} - 1\right]^{-1}.$$

Differentiating each side of this equation with respect to i gives

$$P'(i) = -\$50\left[(1+i)^{\frac{1}{2}} - 1\right]^{-2}\left[\frac{1}{2}(1+i)^{-\frac{1}{2}}\right].$$

It follows that

$$D(i,1) = -\frac{P'(i)}{P(i)} = -\left[\frac{-\$50\left[(1+i)^{\frac{1}{2}}-1\right]^{-2}\left[\frac{1}{2}(1+i)^{-\frac{1}{2}}\right]}{\$50\left[(1+i)^{\frac{1}{2}}-1\right]^{-1}}\right]$$
$$= \frac{\frac{1}{2}(1+i)^{-\frac{1}{2}}}{(1+i)^{\frac{1}{2}}-1},$$

and

$$D(i,\infty) = D(i,1)(1+i) = \frac{\frac{1}{2}(1+i)^{\frac{1}{2}}}{(1+i)^{\frac{1}{2}}-1}.$$

In particular,

$$D(.05,\infty) = \frac{\frac{1}{2}(1.05)^{\frac{1}{2}}}{(1.05)^{\frac{1}{2}}-1} \approx 20.74695077 \approx 20.74695.$$

Moreover, using Important Fact (9.2.19), we obtain

$$D(.05,2) = D(.05,\infty)(1.05)^{-\frac{1}{2}} \approx 20.24695077 \approx 20.24695.$$

(7) We may use the result of Example (9.2.26) with $N = 16$, $m = 1$, and $n = \frac{N}{m} = n$ to obtain a formula for the Macaulay duration; we find that for $i > 0$

$$D(i,\infty) = \frac{1}{d} - \frac{16}{(1+i)^{16}-1}.$$

Therefore, the financial advisor's interest rate i satisfies

$$7.39 = \frac{1+i}{i} - \frac{16}{(1+i)^{16}-1} = 1 + \frac{1}{i} - \frac{16}{(1+i)^{16}-1},$$

and

$$6.39 = \frac{1}{i} - \frac{16}{(1+i)^{16}-1}.$$

We will locate i using the "guess and check" method. Let $f(i) = \frac{1}{i} - \frac{16}{(1+i)^{16}-1}$; we seek i with $f(i) = 6.39$. Note that $f(i) = \frac{(1+i)^{16}-1-16i}{i\left[(1+i)^{16}-1\right]}$ is a rational function, hence is continuous except at those points at which it is undefined due to the denominator being zero; since $i\left[(1+i)^{16}-1\right]$ is non-zero except at $i = 0$ and at $i = -2$, $f(i)$ is continuous for $i > 0$. Therefore, if we determine positive numbers i_1 and i_2 with $f(i_1) < 0$ and $f(i_2) > 0$, there must be a root between i_1 and i_2.

We begin by noting that $f(.04) \approx 6.67200311$, and $f(.08) \approx 5.904625602$, so there is a root between 4% and 8%. We further note that $f(.05) \approx 6.473629447\%$, while $f(6\%) \approx 6.279428376$. At this point, we know there is a root between 5% and 6%. Continuing in this manner, we narrow the interval in which a root is sure to lie. Eventually, we compute $f(5.42791\%) \approx 6.39000096$. Since $f(5.427915\%) \approx 6.389999988$, there is a root between 5.42791% and 5.427915%. Thus, reporting the rate as a percentage correct to five places following the decimal point, our answer is 5.42791%.

It remains for us to show that there was a unique interest rate to be found. For this goal, it is helpful to note that the equation

$$6.39 = \frac{1}{i} - \frac{16}{(1+i)^{16} - 1}$$

is equivalent to each of the following equations:

$$\begin{aligned}
&6.39i\left[(1+i)^{16} - 1\right] = (1+i)^{16} - 1 - 16i;\\
&6.39i(1+i)^{16} - 6.39i = (1+i)^{16} - 1 - 16i;\\
&6.39(1+i)^{17} - 6.39(1+i)^{16} + 9.61i + 1 - (1+i)^{16} = 0;\\
&6.39(1+i)^{17} - 7.39(1+i)^{16} + 9.61(1+i) - 8.61 = 0.
\end{aligned}$$

Therefore, the interest rate that Mr. Linden's financial advisor used was such that $1+i$ was a root of the polynomial

$$p(x) = 6.39x^{17} - 7.39x^{16} + 9.61x - 8.61.$$

Lacking a graphing utility, we need to be a bit clever (*or persistent*) here. We note that $p(1) = 0$, so the polynomial $x - 1$ divides $p(x)$. In fact, $p(x) = (x-1)q(x)$ where

$$q(x) = 6.39x^{16} - x^{15} - x^{14} - x^{13} - x^{12} - \cdots - x^2 - x + 8.61;$$

the polynomial $q(x)$ may be determined by a division of polynomials. [We also note that Equation (3.3.2) may be used to give a simple expression for $q(x)$;

$$q(x) = 6.39x^{16} - x^{15} - x^{14} - x^{13} - x^{12} - \cdots - x^2 - x + 8.61 = 6.39x^{16} - x\left(\frac{x^{15}-1}{x-1}\right) + 8.61.$$

With this expression, the factorization $p(x) = (x-1)q(x)$ is easily verified.]

We next observe that we can factor $p(x)$ further since $q(1) = 0$, and this tells us that the polynomial $x - 1$ divides $q(x)$; therefore $(x-1)^2$ divides $p(x)$. Division of the polynomial $q(x)$ by $(x-1)$ gives $q(x) = (x-1)w(x)$ where

$$\begin{aligned}
w(x) = {}&6.39x^{15} + 5.39x^{14} + 4.39x^{13} + 3.39x^{12} + 2.39x^{11} + 1.39x^{10} + .39x^9\\
&- .61x^8 - 1.61x^7 - 2.61x^6 - 3.61x^5 - 4.51x^4 - 5.61x^3 - 6.61x^2 - 7.61x - 8.61.
\end{aligned}$$

By Descartes' rule of signs, the polynomial $w(x)$ has a unique positive root, and since

$$p(x) = (x-1)q(x) = (x-1)^2 w(x),$$

the polynomial $p(x)$ has only one positive root besides 1. Since we are considering $i > 0$ such that $1+i$ is a root of $p(x)$, we only had one interest rate i to find.

Those of you who have a BA II Plus calculator might note that the equation

$$6.39(1+i)^{17} - 7.39(1+i)^{16} + 9.61(1+i) - 8.61 = 0$$

is also equivalent to

$$-8.61(1+i)^{-17} + 9.61(1+i)^{-16} - 7.39(1+i)^{-1} + 6.39 = 0.$$

So, we are searching for an unknown interest rate consistent with one party paying 6.39 now and 9.61 in sixteen years, in return for 7.39 in one year and 8.61 in seventeen years. This seems like a question that the **Cash Flow worksheet** might solve, but if you enter

CF0=6.39, C01= −7.39, F01=1, C02 = 0, F02=14, C03= 9.61, F03= 1. C04= −8.61, and F04= 1, keying **IRR** **CPT** results in the display "ERROR 7". We remind you that this error message means that the iteration limit is exceeded.

On the other hand, the **Cash Flow Worksheet** can locate the root $(1+i)$ of the polynomial $w(x)$. To accomplish this, enter

CF0 = 6.39, C01 = 5.39, F01=1, C02 = 4.39, F02 = 1, C03=3.39, F03 =1, C04 = 2.39, F04=1,
C05 = 1.39, F05 = 1, C06 = .39, F06 = 1, C07 = −.61, F07 = 1, C08 = −1.61, F08 =1,
C09 = −2.61, F09=1, C10 = −3.61, F10=1, C11=−4.61, F11=1, C12=−5.61, F12=1,
C13 = −6.61, F13=1, C14 = −7.61, F14=1, C15 = −8.61, F15=1,

and then push **IRR** **CPT**. Your display should now read "IRR = 5.427914936". Thus, $i \approx 5.427914936\%$.

(9) Using an annual effective interest rate of 9%, we will calculate the Macaulay duration and the price of the annuity, of the set of coupons from the bond, and of the bond redemption payment. Once we have these six figures, we will use Important Fact (9.2.28) to determine the Macaulay duration of the portfolio.

The annuity may be thought of as repayments of an amortized loan made at interest rate $i = 9\%$. There are $m = 12$ payments per year and $N = 5 \times 12 = 60$ payments remaining. We set $n = \frac{N}{m} = 5$. Then, according to the solution to Example (9.2.26), the Macaulay duration of the annuity is

$$D^{(\text{annuity})}(.09, \infty) = \frac{1}{d^{(12)}} - \frac{5}{(1.09)^5 - 1} \approx 11.64564706 - 9.282914275 \approx 2.362732788.$$

We also note that the price of the annuity to yield the buyer 9% annually (*or loan amount*) is

$$P^{(\text{annuity})} = \$1{,}000a_{\overline{60}|[(1.09)^{\frac{1}{12}}-1]} \approx \$48{,}571.23766.$$

The coupons from the bond form a level annuity with sixteen semiannual payments. So, again using the result found in Example (9.2.26) at $i = 9\%$, the set of all the coupons has Macaulay duration

$$D^{(\text{coupons})}(.09, \infty) = \frac{1}{d^{(2)}} - \frac{8}{(1.09)^8 - 1} \approx 11.85572584 - 8.059944697 \approx 3.795781142.$$

The amount of each coupon is $\$20{,}000\left(\frac{.08}{2}\right) = \$800$, so the price of the annuity of coupons, which yields the buyer 9%, is

$$P^{(\text{coupons})} = \$800a_{\overline{16}|[(1.09)^{\frac{1}{2}}-1]} \approx \$9{,}050.671934.$$

The set consisting of just the \$20,000 redemption payment eight years into the future has Macaulay duration

$$D^{(\text{redem})}(.09, \infty) = 8.$$

Its price, to yield $i = 9\%$, is

$$P^{(\text{redem})} = \$20{,}000(1.09)^{-8} \approx \$10{,}037.32559.$$

Therefore, the total price of the portfolio is

$$P = P^{(\text{annuity})} + P^{(\text{coupons})} + P^{(\text{redem})} \approx \$48{,}571.23766 + \$9{,}050.671934 + \$10{,}037.32559 \approx \$67{,}659.23519.$$

Important Fact (9.2.28) tells us that the duration of the portfolio is a weighted average of the three durations we have found, the weights given to the durations being

$$\frac{P^{(\text{annuity})}}{P} \approx .717880383, \quad \frac{P^{(\text{coupons})}}{P} \approx .133768464, \text{ and } \frac{P^{(\text{redem})}}{P} \approx .148351154,$$

respectively. In fact, the Macaulay duration of the portfolio $D(,09,\infty)$ satisfies

$$D(,09,\infty) = \left(\frac{P^{(\text{annuity})}}{P}\right)\left[D^{(\text{annuity})}(.09,\infty)\right] + \left(\frac{P^{(\text{coupons})}}{P}\right)\left[D^{(\text{coupons})}(.09,\infty)\right] + \left(\frac{P^{(\text{redem})}}{P}\right)\left[D^{(\text{redem})}(.09,\infty)\right]$$
$$\approx (.717880383)(2.362732788) + (.133768464)(3.795781142) + (.148351154)(8)$$
$$\approx 3.390724558 \approx 3.39072.$$

(9.3) Convexity

(1) As in problem (9.2.1), we have $F = \$1{,}000$, $r = 6\%$, and $C = \$1{,}200$. So, the bond pays \$60 at times $t = 1, 2, \ldots, 9$ and $\$60 + \$1{,}200 = \$1{,}260$ at $t = 10$. Moreover, the price of the bond, to yield 8%, is

$$P(8\%) = \$60a_{\overline{10}|8\%} + \$1{,}200(1.08)^{-10} \approx \$958.437069.$$

According to Important Fact (9.3.8), the Macaulay convexity $C(8\%,\infty)$ is

$$C(8\%,\infty) = \left(\sum_{k=1}^{9}\left[\frac{\$60(1.08)^{-k}}{P(8\%)}\right]k^2\right) + \left[\frac{\$1{,}260(1.08)^{-10}}{P(8\%)}\right](10)^2$$
$$= \frac{1}{P(8\%)}\left[\sum_{k=1}^{9}\$60k^2(1.08)^{-k} + \$126{,}000(1.08)^{-10}\right]$$
$$\approx \frac{1}{958.437069}[68{,}361.33989]$$
$$\approx 71.32585128 \approx 71.32585.$$

We note that the numerical value for the sum

$$\sum_{k=1}^{9} 60k^2(1.08)^{-k} + 126{,}000(1.08)^{-10}$$

may be calculated using the **Cash Flow worksheet** with
CF0=0, C01= $60(1^2) = 60$, C02=$60(2^2) = 240$, C03=$60(3^2) = 540$, C04= $60(4^2) = 960$, C05=$60(5^2) = 1{,}500$, C06=$60(6^2) = 2{,}160$, C07 = $60(7^2) = 2{,}940$, C08 =$60(8^2) = 3{,}840$, C09 =$60(9^2) = 4{,}860$, C10= $1{,}260(10^2)$, and F01= F02 = F03 = ... =F10 =1.
With these entries made, press **NPV** **8** **ENTER** **↓** **CPT** to display "NPV=68,361.33989".

(3) Applying Equation (9.3.9) with $m = 4$, we have

$$C(i,4) = \frac{C(i,\infty) + \frac{1}{4}D(i,\infty)}{\left(1 + \frac{i^{(4)}}{4}\right)^2}.$$

Therefore, since $1 + i = \left(1 + \frac{i^{(4)}}{4}\right)^4$, and we are given $D(i,\infty) = 5.8$ and $C(i,\infty) = 1.2$,

$$C(i,4) = \frac{1.2 + \frac{5.8}{4}}{\left(1 + \frac{i^{(4)}}{4}\right)^2} = \frac{2.65}{(1+i)^{\frac{1}{2}}} = \frac{2.65}{\sqrt{1+i}}.$$

(5) (a) Recall that when accumulation is based on compound interest at an annual effective interest rate i, the force of interest is $\delta = \ln(1+i)$. Since $1+i = \left(1+\frac{i^{(m)}}{m}\right)^m$, this gives us

$$\delta = \ln\left(1+\frac{i^{(m)}}{m}\right)^m = m\ln\left(1+\frac{i^{(m)}}{m}\right).$$

Differentiating with respect to $i^{(m)}$ and remembering to use the chain rule, we obtain

$$\frac{d\delta}{di^{(m)}} = m\left(\frac{1}{1+\frac{i^{(m)}}{m}}\right)\cdot\frac{1}{m} = \frac{1}{1+\frac{i^{(m)}}{m}} = \left(1+\frac{i^{(m)}}{m}\right)^{-1}.$$

(b) The result is obtained by taking the derivative of the equation found in part (a), once again taking care to remember to use the chain rule. Specifically, we have

$$\frac{d^2\delta}{d(i^{(m)})^2} = \frac{d}{di^{(m)}}\left[\frac{d\delta}{di^{(m)}}\right] = \frac{d}{di^{(m)}}\left[\left(1+\frac{i^{(m)}}{m}\right)^{-1}\right] = -1\left(1+\frac{i^{(m)}}{m}\right)^{-2}\cdot\frac{1}{m} = -\frac{1}{m}\left(1+\frac{i^{(m)}}{m}\right)^{-2}.$$

(c) Using the chain rule, we find

$$\frac{d^2P}{d(i^{(m)})^2} = \frac{d}{di^{(m)}}\left(\frac{dP}{d(i^{(m)})}\right) = \frac{d}{di^{(m)}}\left(\frac{dP}{d\delta}\cdot\frac{d\delta}{di^{(m)}}\right).$$

To continue, we use the product rule; this gives

$$\frac{d^2P}{d(i^{(m)})^2} = \left[\frac{d}{di^{(m)}}\left(\frac{dP}{d\delta}\right)\right]\cdot\frac{d\delta}{di^{(m)}} + \frac{dP}{d\delta}\cdot\frac{d^2\delta}{d(i^{(m)})^2}.$$

(d) According to the chain rule,
$\frac{dW}{di^{(m)}} = \frac{dW}{d\delta}\cdot\frac{d\delta}{di^{(m)}}$. In particular, setting $W = \frac{dP}{d\delta}$, we have

$$\frac{d}{di^{(m)}}\left(\frac{dP}{d\delta}\right) = \frac{d^2P}{d\delta^2}\cdot\frac{d\delta}{di^{(m)}}.$$

Recalling the result of part (c), we have

$$\begin{aligned}\frac{d^2P}{d(i^{(m)})^2} &= \left(\left[\frac{d}{di^{(m)}}\left(\frac{dP}{d\delta}\right)\right]\cdot\frac{d\delta}{di^{(m)}}\right) + \left(\frac{dP}{d\delta}\cdot\frac{d^2\delta}{d(i^{(m)})^2}\right)\\ &= \left(\left[\frac{d^2P}{d\delta^2}\cdot\frac{d\delta}{di^{(m)}}\right]\cdot\frac{d\delta}{di^{(m)}}\right) + \left(\frac{dP}{d\delta}\cdot\frac{d^2\delta}{d(i^{(m)})^2}\right)\\ &= \left(\frac{d^2P}{d\delta^2}\cdot\frac{d^2\delta}{d^2i^{(m)}}\right) + \left(\frac{dP}{d\delta}\cdot\frac{d^2\delta}{d(i^{(m)})^2}\right).\end{aligned}$$

Substituting $\left(1+\frac{i^{(m)}}{m}\right)^{-1}$ for $\frac{d\delta}{di^{(m)}}$ [from part(a)] and $-\frac{1}{m}\left(1+\frac{i^{(m)}}{m}\right)^{-2}$ for $\frac{d^2\delta}{d(i^{(m)})^2}$ [from part(b)], we obtain

$$\frac{d^2P}{d(i^{(m)})^2} = \frac{d^2P}{d\delta^2} \cdot \left(1+\frac{i^{(m)}}{m}\right)^{-2} + \frac{dP}{d\delta} \cdot \left[-\frac{1}{m}\left(1+\frac{i^{(m)}}{m}\right)^{-2}\right]$$
$$= \frac{d^2P}{d\delta^2} \cdot \left(1+\frac{i^{(m)}}{m}\right)^{-2} - \frac{dP}{d\delta} \cdot \frac{1}{m}\left(1+\frac{i^{(m)}}{m}\right)^{-2}.$$

(e) According to Equation (9.3.5), $C(i,m) = \frac{d^2P}{di^{(m)2}} \Big/ P(i)$, and Equation (9.3.6) states that $C(i,\infty) = \frac{d^2P}{d\delta^2} \Big/ P(i)$. If we divide the result of part (d) by the price function $P(i)$, we have

$$\frac{d^2P}{d(i^{(m)})^2} \Big/ P(i) = \left(\frac{d^2P}{d\delta^2} \Big/ P(i)\right) \cdot \left(1+\frac{i^{(m)}}{m}\right)^{-2} - \left(\frac{1}{m}\left[-\frac{dP}{d\delta} \Big/ P(i)\right]\right) \cdot \left(1+\frac{i^{(m)}}{m}\right)^{-2}.$$

But, according to Equation (9.3.5), $C(i,m) = \frac{d^2P}{di^{(m)2}} \Big/ P(i)$, and Equation (9.3.6) states that $C(i,\infty) = \frac{d^2P}{d\delta^2} \Big/ P(i)$. Therefore, our last result is equivalent to

$$C(i,m) = \left(C(i,\infty) + \frac{1}{m}D(i,\infty)\right) \Big/ \left(1+\frac{i^{(m)}}{m}\right)^2.$$

(9.4) Immunization

(1) Let P_1 and P_4 denote the prices of the one-year bond and of the four-year bond, respectively, so that the present value and duration conditions of Redington immunization are satisfied. Then the combined assets have present value $P_A = P_1 + P_4$. Moreover, using Important Fact (9.2.21), the set of assets has Macaulay duration

$$D_A = \left(\frac{P_1}{P_A}\right)1 + \left(\frac{P_4}{P_A}\right)4 = \left(\frac{P_1}{P_1+P_4}\right)1 + \left(\frac{P_4}{P_1+P_4}\right)4.$$

Since the liability has present value $P_L = \$300{,}000(1.04)^{-3}$ and Macaulay duration $D_L = 3$, we are looking for P_1 and P_4 satisfying the system of simultaneous equations

$$\begin{cases} (1)\quad P_1 + P_4 = P_A = P_L = \$300{,}000(1.04)^{-3}, \\ (2)\quad \left(\frac{P_1}{P_1+P_4}\right)1 + \left(\frac{P_4}{P_1+P_4}\right)4 = D_A = D_L = 3. \end{cases}$$

Multiplying Equation (2) by $(P_1 + P_4)$, we obtain the equivalent equation

$$P_1 + 4P_4 = 3P_1 + 3P_4,$$

and this in turn is equivalent to the equation $P_4 = 2P_1$. Thus, we may replace Equation (2) with the equation $P_4 = 2P_1$. The linear system

$$\begin{cases} P_1 + P_4 = \$300{,}000(1.04)^{-3}, \\ P_4 = 2P_1 \end{cases}$$

has the unique solution

$$P_1 = \$100{,}000(1.04)^{-3} \approx \$88{,}899.64,\ P_4 = \$200{,}000(1.04)^{-3} \approx \$177{,}799.27.$$

Note that the liability has convexity $C_L = 3^2 = 9$, and using Important Fact (9.3.8), the assets have convexity

$$C_A = \left(\frac{P_1}{P_1 + P_4}\right) 1^2 + \left(\frac{P_4}{P_1 + P_4}\right) 4^2 = \left(\frac{1}{3}\right) 1^2 + \left(\frac{2}{3}\right) 4^2 = 11.$$

So, the convexity condition $C_A > C_L$ is satisfied.

(3) Based on a 4.5% annual effective interest rate, the present value of the liability is $\$500{,}000(1.045)^{-4} \approx$ $\$419{,}280.67$, while the present value of the assets is $P_A = \$343{,}398.73(1.045)^{-2} + \$162{,}782.52(1.045)^{-10} \approx$ $\$419{,}280.67$. Therefore, the present value condition is said to be satisfied. The Macaulay duration of the liability is the lone payment time; $D_L = 4$. The Macaulay duration of the assets is a price-weighted average of the cashflow times;

$$D_A = \left(\frac{\$343{,}398.73(1.045)^{-2}}{P_A}\right) 2 + \left(\frac{\$162{,}782.52(1.045)^{-10}}{P_A}\right) 10$$
$$\approx 4.000000048 \approx 4.$$

Therefore, the duration condition is also judged to be satisfied, and we say that the portfolio is "fully immunized". If interest rates immediately rise to 8%, then the surplus will be

$$S(8\%) = \$343{,}398.73(1.08)^{-2} + \$162{,}782.52(1.08)^{-10} - \$500{,}000(1.08)^{-10} \approx \$2{,}293.938911 \approx \$2{,}293.94.$$

In one year, using an annual effective interest rate of $i = 4.5\%$, the present values have all been multiplied by a factor of 1.045, so the present values of the set of assets still "equals" the present value of the liabilities. Moreover, the Macaulay duration of the liability has decreased by 1, as has the Macaulay duration of the set of assets, so we still have full immunization. At this later time, if the interest rate suddenly jumps to 8%, the surplus will be 1.08 times the previously calculated surplus; it is approximately \$2,477.45.

(9.5) Other types of duration

(1) First observe that if we denote the current interest rate by i_0, then we were given $P(i_0) = 95.40$, $P(i_0 + .01) = 92.50$, and $P(i_0 - .01) = 96.60$. Therefore, by Equation (9.5.1),

$$m_{.01} = \frac{92.50 - 96.60}{2(.01)} = -205.$$

Equation (9.5.2) then gives us the effective duration;

$$E_{.01}(i_0, 1) = -\frac{m_{.01}}{P(i_0)} = -\frac{205}{95.40} \approx 2.14884696 \approx 2.14885.$$

(3) The price of the bond is given to be \$49,200 if it is to yield the purchaser an annual effective yield of 5%, \$48,392.75 if it is to yield an annual effective rate of 6%, and \$47,181.90 if it is to yield an annual effective interest rate of 7%. Therefore,

$$E_{.01}(.06, 1) = -\frac{m_{.01}}{P(.06)} = -\left(\frac{\$47{,}181.90 - \$49{,}200}{2(.01)} \Big/ \$48{,}392.75\right) \approx 2.0851263088 \approx 2.08513.$$

If the annual effective yield rate falls from 6% to 5.4%, a drop of $.06 - .054 = .006$, then since

$$(.006)[E_{.01}(.06,1)] \approx (.006)(2.0851263088) \approx .012510758,$$

we estimate the price to rise to

$$(1.012510758)(\$48{,}392.75) = \$48{,}998.18.$$

If we have a larger drop of the annual effective yield rate falls from 6% to 4%, a decrease of $.06 - .04 = .02$, then we estimate that the price of the bond rises to

$$\big(1 + (.02)[E_{.01}(.06,1)]\big)(\$48{,}392.75) = \$50{,}410.85.$$

We note that if not for the call feature, the price of the bond at annual effective interest rate i would be $\$2{,}500(1+i)^{-1} + \$2{,}500(1+i)^{-2} + \$54{,}200(1+i)^{-3}$; so, at $i = 5.4\%$, it would be approximately

$$\$2{,}500(1.054)^{-1} + \$2{,}500(1.054)^{-2} + \$54{,}200(1.054)^{-3} \approx \$50{,}911.27,$$

and at 4%, the price of the noncallable bond would be

$$\$2{,}500(1.04)^{-1} + \$2{,}500(1.04)^{-2} + \$54{,}200(1.04)^{-3} \approx \$52{,}898.84.$$

If the bond is noncallable, its Macaulay duration $D(.06,\infty)$ is given as a price-weighted average of its three payments. Note that when determined using the interest rate $i = 6\%$, the present value of the payment at time 1 is $P_1 = \$2{,}500(1.06)^{-1}$, the present value of the payment at time 2 is $P_2 = \$2{,}500(1.06)^{-2}$, and the present value of the payment at time 3 is $P_3 = \$54{,}200(1.06)^{-3}$. Thus,

$$\begin{aligned} D(.06,\infty) &= \left(\frac{P_1}{P_1+P_2+P_3}\right)1 + \left(\frac{P_2}{P_1+P_2+P_3}\right)2 + \left(\frac{P_3}{P_1+P_2+P_3}\right)3 \\ &\approx \left(\frac{\$2{,}358.490566}{\$50{,}090.84681}\right)1 + \left(\frac{\$2{,}224.9911}{\$50{,}090.84681}\right)2 + \left(\frac{\$45{,}507.36514}{\$50{,}090.84681}\right)3 \\ &\approx 2.86141236. \end{aligned}$$

Consequently,

$$D(.06,1) = D(.06,\infty)/(1.06) \approx 2.86141236/1.06 \approx 2.699445623.$$

Moreover, the price of the noncallable bond, to yield the buyer 6% annually, is

$$P_1 + P_2 + P_3 \approx \$50{,}090.84681 \approx \$50{,}090.85022 \approx \$50{,}090.85.$$

Using this price, along with the value we just calculated for $D(.06,1)$, we find that if the price falls to 5.4%, the estimated price for the noncallable bond is

$$P(5.4\%) \approx \big(1 + (.006)[D(.06,1)]\big)(\$50{,}090.84681) \approx \$50{,}902.15191 \approx \$50{,}902.15$$

Moreover, if the price drops to 4%, our estimate for the price of the noncallable bond is

$$P(4\%) \approx \big(1 + (.02)[D(.06,1)]\big)(\$50{,}090.84681) \approx \$52{,}795.19715 \approx \$52{,}795.20.$$

These estimated prices are a bit higher than the actual prices calculated above.

The callable bond's estimated prices are lower than the estimated prices for the noncallable bond. This is as it should be, since the holder of the callable bond faces the possible disadvantages of having the bond called and is not compensated for this by call premiums.

(5) The bond has $m = 1$, $n = N = 8$, and $r = \alpha = 1$. Moreover, its current price (to yield $i_0 = 6.6\%$ annually) is $P(6.6\%) = \$102.43$. In general, the effective duration is

$$E_h(i_0, 1) = -\frac{P(i_0 + h) - P(i_0 - h)}{2h} \Big/ P(i_0),$$

while the effective convexity is

$$F_h(i_0, 1) = \frac{P(i_0 + h) + P(i_0 - h) - 2P(i_0)}{h^2} \Big/ P(i_0).$$

Since $6.8\% - 6.6\% = .2\% = .002$ and $6.4\% - 6.6\% = .2\% = .002$, the relevant h here is $h = .002$. We calculate

$$\begin{aligned} E_{.002}(i_0, 1) &= -\frac{P(6.8\%) - P(6.4\%)}{2(.002)} \Big/ P(6.6\%) \\ &= -\frac{\$101.20 - \$103.21}{.004} \Big/ \$102.43 \\ &\approx 4.90578932 \approx 4.90579, \end{aligned}$$

and

$$\begin{aligned} F_{.002}(i_0, 1) &= \frac{P(6.8\%) + P(6.4\%) - 2P(6.6\%)}{2(.002)} \Big/ P(6.6\%) \\ &= \frac{\$101.20 + \$103.21 - 2(\$102.43)}{.004} \Big/ \$102.43 \\ &\approx -1{,}098.311042 \approx -1{,}098.31104. \end{aligned}$$

Recall that the second Taylor polynomial approximation

$$P(i) \approx P(i_0) + P'(i_0)(i - i_0) + \frac{P''(i_0)}{2}(i - i_0)^2$$

gave rise to the approximation [Equation(9.3.3)]

$$\frac{P(i) - P(i_0)}{P(i_0)} \approx -D(i_0, 1)(i - i_0) + C(i_0, 1)\frac{(i - i_0)^2}{2}.$$

Now, we substitute the effective duration for the modified duration and the effective duration for the effective convexity; we find

$$\frac{P(i) - P(i_0)}{P(i_0)} \approx -E_h(i_0, 1)(i - i_0) + F_h(i_0, 1)\frac{(i - i_0)^2}{2}.$$

In particular, if $i_0 = 6.6\%$ and $i = 7.2\%$, we have

$$\begin{aligned} \frac{P(7.2\%) - P(6.6\%)}{P(6.6\%)} &\approx -(4.90578932)(.072 - .066) + (-1{,}098.311042)\frac{(.072 - .066)^2}{2} \\ &\approx -.049204335. \end{aligned}$$

Therefore,

$$P(7.2\%) \approx (1 - .049204335)P(6.6\%) \approx (.950795665)(\$102.43) \approx \$97.39.$$

Chapter 9 review problems

(1) We are considering a bond with $F = C$, $m = 1$, $N = n = 15$, and $r = \alpha = 7.5\%$. If a buyer purchases the bond for the redemption amount, then the yield will be the coupon rate 7.5%. Therefore, $D(7.5\%, 1)$ may be found using the result of Example (9.2.24); we obtain

$$D(7.5\%, 1) = a_{\overline{15}|7.5\%} \approx 8.827119745 \approx 8.82712.$$

The bond pays fifteen annual coupons, each for an amount $(.075)F$, along with the redemption payment of F in fifteen years. Therefore, for any annual effective interest rate i, we have

$$P(i) = \left[\sum_{k=1}^{14}(.075F)(1+i)^{-k}\right] + (1.075F)(1+i)^{-15},$$

and

$$\begin{aligned} P'(i) &= \left[\sum_{k=1}^{14}(-k)(.075F)(1+i)^{-(k+1)}\right] + (-15)(1.075F)(1+i)^{-16} \\ &= -F\left(.075\left[(1(1+i)^{-2} + 2(1+i)^{-3} + \cdots + 14(1+i)^{-15}\right] + 16.125(1+i)^{-16}\right) \\ &= -F\left(.075(1+i)^{-1}\left[(1(1+i)^{-1} + 2(1+i)^{-2} + \cdots + 14(1+i)^{-14}\right] + 16.125(1+i)^{-16}\right) \\ &= -F\left(.075(1+i)^{-1}(Ia)_{\overline{14}|i} + 16.125(1+i)^{-16}\right) \\ &= -F\left(.075(1+i)^{-1}\left[\frac{\ddot{a}_{\overline{14}|i} - 14(1+i)^{-14}}{i}\right] + 16.125(1+i)^{-16}\right). \end{aligned}$$

Since $D(i,1) = -\frac{P'(i)}{P(i)}$, this gives us a method for calculating $D(i,1)$ for an arbitrary i. If you wish to check the value of $D(7.5\%, 1)$ that we found above, note that the above equations give

$$P'(7.5\%) \approx -F(8.827119745) \quad \text{and} \quad P'(7.5\%) = F.$$

So,

$$D(7.5\%, 1) = -\frac{P'(7.5\%)}{P(7.5\%)} \approx \frac{8.827119745F}{F} = 8.827119745 \approx 8.82712.$$

Again using the above formulas, we obtain

$$P(9\%) \approx F(.879089674) \quad \text{and} \quad P'(9\%) \approx -F(7.34614765).$$

It follows that

$$D(9\%, 1) = -\frac{P'(7.5\%)}{P(7.5\%)} \approx 8.8357412203 \approx 8.83574.$$

(3) We first determine the Macaulay duration and the Macaulay convexity of each of the bonds, using a superscript to indicate the term of the bond; for example , we denote the Macaulay duration of the two-year bond by $D^{(2)}(5\%, \infty)$. Finding the Macaulay duration and the Macaulay convexity for the two zero-coupon bonds is especially easy ; we have

$$D^{(2)}(5\%, \infty) = 2, \ , D^{(3)}(5\%, \infty) = 3, \ C^{(2)}(5\%, \infty) = 2^2 = 4, \text{ and } C^{(3)}(5\%, \infty) = 3^2 = 9.$$

Let $P^{(5)}(5\%)$ denote the price of the five year coupon bond; by the basic price formula,

$$P^{(5)}(5\%) = \$60a_{\overline{5}|5\%} + \$1{,}000(1.05)^{-5} \approx \$1{,}043.294767.$$

Using Important Fact (9.2.21), the Macaulay duration of the five year coupon bond is

$$\begin{aligned} D^{(5)}(5\%,\infty) &= \left[\frac{60(1+i)^{-1}}{P^{(5)}(5\%)}\right]1+\left[\frac{60(1+i)^{-2}}{P^{(5)}(5\%)}\right]2+\left[\frac{60(1+i)^{-3}}{P^{(5)}(5\%)}\right]3+\left[\frac{60(1+i)^{-4}}{P^{(5)}(5\%)}\right]4+\left[\frac{1{,}060(1+i)^{-5}}{P^{(5)}(5\%)}\right]5 \\ &\approx 4.477751243 \approx 4.47775. \end{aligned}$$

Important Fact (9.3.8) tells us that

$$\begin{aligned} C^{(5)}(5\%,\infty) &= \left[\frac{60(1+i)^{-1}}{P^{(5)}(5\%)}\right]1^2+\left[\frac{60(1+i)^{-2}}{P^{(5)}(5\%)}\right]2^2+\left[\frac{60(1+i)^{-3}}{P^{(5)}(5\%)}\right]3^2+\left[\frac{60(1+i)^{-4}}{P^{(5)}(5\%)}\right]4^2+\left[\frac{1{,}060(1+i)^{-5}}{P^{(5)}(5\%)}\right]5^2 \\ &\approx 21.36935863 \approx 21.36936. \end{aligned}$$

The price $P(5\%)$ of the portfolio to yield 5% is the sum of the prices of the individual bonds. Since the two-year and three-year bonds are each zero-coupon \$1,000 bonds, the two-year bond has price $P^{(2)}(5\%) = \$1{,}000(1.05)^{-2} \approx \$952.3809524$, and the three-year bond has price $P^{(3)}(5\%) = \$1{,}000(1.05)^{-3} \approx \$907.0294785$. We have already noted that $P^{(5)}(5\%) \approx \$1{,}043.294767$. Thus,

$$\begin{aligned} P(5\%) &= P^{(2)}(5\%) + P^{(3)}(5\%) + P^{(5)}(5\%) \\ &\approx \$907.0294785 + \$952.3809524 + \$1{,}043.294767 \\ &\approx \$2{,}882.539032. \end{aligned}$$

To calculate the Macaulay duration and the Macaulay convexity of the portfolio, we use Important Facts (9.2.28) and (9.3.16); these allow us to find the portfolio Macaulay duration and the Macaulay convexity as price-weighted averages of the individual Macaulay durations and the Macaulay convexities. Specifically, we find that the portfolio had Macaulay duration

$$\begin{aligned} D^{\text{portfolio}}(5\%,\infty) &= \left[\frac{P^{(2)}(5\%)}{P(5\%)}\right]D^{(2)}(5\%,\infty)+\left[\frac{P^{(3)}(5\%)}{P(5\%)}\right]D^{(3)}(5\%,\infty)+\left[\frac{P^{(5)}(5\%)}{P(5\%)}\right]D^{(5)}(5\%,\infty) \\ &= \frac{1}{P(5\%)}\left[P^{(2)}(5\%)D^{(2)}(5\%,\infty)+P^{(3)}(5\%)D^{(3)}(5\%,\infty)+P^{(5)}(5\%)D^{(5)}(5\%,\infty)\right] \\ &\approx \frac{1}{2{,}882.539032}\left[(907.0294785)(2)+(952.3809524)(3)+(1{,}043.294767)(4.477751243)\right] \\ &\approx 3.22553808 \approx 3.22554, \end{aligned}$$

and Macaulay convexity

$$\begin{aligned} C^{\text{portfolio}}(5\%,\infty) &= \left[\frac{P^{(2)}(5\%)}{P(5\%)}\right]C^{(2)}(5\%,\infty)+\left[\frac{P^{(3)}(5\%)}{P(5\%)}\right]D^{(3)}(5\%,\infty)+\left[\frac{P^{(5)}(5\%)}{P(5\%)}\right]C^{(5)}(5\%,\infty) \\ &= \frac{1}{P(5\%)}\left[P^{(2)}(5\%)C^{(2)}(5\%,\infty)+P^{(3)}(5\%)C^{(3)}(5\%,\infty)+P^{(5)}(5\%)C^{(5)}(5\%,\infty)\right] \\ &\approx \frac{1}{2{,}882.539032}\left[(907.0294785)(4)+(952.3809524)(9)+(1{,}043.294767)(21.36935863)\right] \\ &\approx 11.9741501 \approx 11.97415. \end{aligned}$$

The **Cash Flow worksheet** and **NPV subworksheet** can save you some work in the calculations of this problem.

(5) First consider the liabilities. There is a single liability of \$250,000 due in five years, and figured using an annual effective interest rate of 5%, this has price (*present value*)

$$P_L = \$250{,}000(1.05)^{-5} \approx \$195{,}881.5416.$$

Since the liabilities consist of a single payment in five years, the Macaulay duration of the liabilities is $D_L = 5$.

The assets are to consist of two zero-coupon bonds, with face values yet to be determined, and we let P_2 and P_7 denote their respective prices to yield the buyer 5% annually; here, the subscripts give the term of the bond. So, the total price of the assets is $P_A = P_2 + P_7$. Applying Important Fact (9.2.21) , the assets have Macaulay duration

$$D_A = \left(\frac{P_2}{P_A}\right)2 + \left(\frac{P_2}{P_A}\right)7 = \left(\frac{P_2}{P_2+P_7}\right)2 + \left(\frac{P_2}{P_2+P_7}\right)7.$$

National Reliance Insurance chooses P_2 and P_7 so their portfolio of assets and liabilities will be immunized. Therefore, they seek P_2 and P_7 so that Important Fact (9.4.6) will apply; that is, we look for P_2 and P_7 so that $P_L = P_A$ and $D_L = D_A$; otherwise put, we wish to solve the system of simultaneous equations

$$\begin{cases} (1)\quad P_2 + P_7 = P_L = \$250{,}000(1.05)^{-5}, \\ (2)\quad \left(\frac{P_2}{P_2+P_7}\right)2 + \left(\frac{P_2}{P_2+P_7}\right)7 = D_A = D_L = 5. \end{cases}$$

Multiplying Equation (2) by $(P_2 + P_7)$, we see that Equation (2) is equivalent to

$$2P_2 + 7P_7 = 5P_2 + 5P_7.$$

Consequently, Equation (2) is equivalent to $2P_7 = 3P_2$ and to

$$P_2 = \frac{2}{3}P_7.$$

Substituting $\frac{2}{3}P_7$ for P_2 in Equation (1), we have

$$\frac{2}{3}P_7 + P_7 = \$250{,}000(1.05)^{-5}.$$

It follows that

$$P_7 = \frac{3}{5}\left[\$250{,}000(1.05)^{-5}\right] \approx \$117{,}528.925,$$

and

$$P_2 = \frac{2}{5}\left[\$250{,}000(1.05)^{-5}\right] \approx \$78{,}352.61665.$$

Therefore, we take

$$P_2 = \$78{,}352.62 \quad \text{and} \quad P_7 = \$117{,}528.92.$$

Since $\$78{,}352.62(1.05)^2 \approx \$86{,}383.76355$, the two-year bond has redemption amount \$86,383.76. Moreover, $\$117{,}528.92(1.05)^7 \approx \$165{,}374.993$, so the seven year bond has redemption amount 165,374.99. Thus, if the annual effective interest rate immediately drops to 4%, then the present value of the portfolio is

$$\$86{,}383.76(1.04)^{-2} + 165{,}374.99(1.04)^{-7} - \$250{,}000(1.04)^{-5} \approx \$56.266081 \approx \$56.27.$$